MECHANISMS OF LYMPHOCYTE ACTIVATION AND IMMUNE REGULATION VIII

ADVANCES IN EXPERIMENTAL MEDICINE AND BIOLOGY

Recent Volumes in this Series

MECHANISMS OF LYMPHOCYTE ACTIVATION AND IMMUNE REGULATION VIII
AUTOIMMUNITY 2000 AND BEYOND

Edited by

Sudhir Gupta
University of California, Irvine
Irvine, California

Springer Science+Business Media, LLC

Library of Congress Cataloging-in-Publication Data

Mechanisms of lymphocyte activation and immune regulation VIII: autoimmunity 2000
and beyond/edited by Sudhir Gupta.
     p. ; cm.—(Advances in experimental medicine and biology; v. 490)
     Includes bibliographical references and index.
     ISBN 978-1-4613-5458-1   ISBN 978-1-4615-1243-1 (eBook)
     DOI 10.1007/978-1-4615-1243-1
     1. Autoimmunity. 2. Autoimmune diseases—Pathogenesis. I. Title: Autoimmunity
2000 and beyond. II. Gupta, Sudhir. III. Series.
     [DNLM: 1. Autoimmunity. 2. Autoimmune Diseases—therapy. 3. Lymphocyte
Transformation. 4. Lymphocytes—immunology. QW 545 M486 2001]
     QR188.3 .M43 2001
     616.97′7—dc21

                   2001018616

ISBN 978-1-4613-5458-1

©2001 Springer Science+Business Media New York
Originally published by Kluwer Academic / Plenum Publishers, New York in 2001
Softcover reprint of the hardcover 1st edition 2001

http://www.wkap.nl/

10 9 8 7 6 5 4 3 2 1

A C.I.P. record for this book is available from the Library of Congress

PREFACE

Advances in biochemistry, cell biology, genome-wide mutagenesis—coupled with molecular technology, including gene microarray and transgenic and knock out animals—have been instrumental in understanding the cellular processes and molecular pathways of self-tolerance and autoimmune diseases. The molecular definition of these pathways and processes has lead to novel treatments for certain autoimmune diseases that are based on the pathogenesis of disease rather than on broad-spectrum immunosuppression. This book reviews many of these current developments and proposes future novel approaches for understanding the pathogenesis of autoimmune diseases and designing novel therapy. This book covers three major areas of autoimmunity: the basic mechanisms of immunological tolerance, pathogenesis of autoimmune diseases, and some novel therapies.

Immunological and genetic characterization of mouse models that spontaneously develop lupus has provided important insight into the pathogenesis of autoimmune disease. Recent studies of genomic scan of certain congenic strains of experimental animals have identified genomic intervals containing disease susceptibility loci (e.g. *Sle1*, *Sle2*, *Sl3*, *C1q*, *Sap*). Cellular and molecular pathways and genes that underpin immunological self-tolerance are beginning to be mapped. The role of activated APC, co-stimulatory molecules, regulatory T cells (via cytokines and cytokine receptors), death receptors, and pro-survival molecules in the inhibition of the induction of tolerance and the activation of autoreactive lymphocytes has been defined. The significance of immunoregulatory T cells (through cytokines) in immunological tolerance is evident in the neonatal and adult models of autoimmune gastritis. Experimental models of spontaneous autoimmune diseases and human autoimmune lymphoproliferative disorders have highlighted a role of death receptors, their ligands, and upstream caspases in immunological tolerance. Data for the role of MHC class II molecules and inhibitors of cytokine receptors in type I diabetes mellitus have been presented and discussed. A model of the functional role of epitope spreading in the pathogenesis of chronic autoimmune and virus-induced demyelinating diseases that mimic multiple sclerosis has been reviewed. The use of gene microarray has revealed the differences between EAE and multiple sclerosis. The pros and cons of using EAE as a predictor of success of new drugs in multiple sclerosis are discussed. Novel treatments of autoimmunity in a lupus model, using co-stimulatory molecules and anti-TNF antibody, are reviewed.

This book should be useful for immunologists, molecular biologists, rheumatologists, and clinician scientists.

I thank Miss Nancy Doman for editorial assistance.

Sudhir Gupta

CONTENTS

THERAPY

DELINEATION OF THE PATHOGENESIS OF SYSTEMIC LUPUS ERYTHEMATOSUS BY USING MURINE MODELS

Kui Liu and Edward K. Wakeland

Center for Immunology
The University of Texas Southwestern Medical Center at Dallas
5323 Harry Hines Blvd.
Dallas, TX 75235-9093

Systemic lupus erythematosus (SLE) is an autoimmune disease that is characterized by the production of autoantibodies to a spectrum of nuclear antigens, and the development of inflammatory processes potentially affecting multiple organ systems with varied clinical manifestations.[1,2] Although the etiology of SLE is still unclear, it has been established that genetic factors play an important role in determining disease susceptibility.[3]

The immunologic and genetic characterization of mouse models that spontaneously develop SLE has provided important insights into the pathogenesis of the disease.[2,4] These mouse models fall into two categories. The first category includes the classic lupus-prone mouse strains, such as (NZB x NZW)F1, BXSB, and MRL-*lpr/lpr*. The second category includes synthetic models produced via the targeted disruption of specific genes, such as C1q and Sap "knockout" mice. These mouse strains also produce high levels of antinuclear antibodies (ANAs) and develop lupus-like glomerulonephritis. Genetic analyses of lupus-prone strains have led to the mapping and characterization of multiple lupus susceptibility loci. We anticipate that detailed characterizations of pathogenic mechanisms mediating disease in these models will provide important insights into the molecules and pathways that maintain tolerance, and identify immune dysregulations involved in the development of autoimmunity. These studies may also provide a better understand of the mechanisms contributing to the pathogenesis of human SLE.

1. MULTIPLE SUSCEPTIBILITY LOCI IDENTIFIED BY LINKAGE ANALYSES

Extensive genetic linkage analyses[5,6,7,8,9,10,11] have identified multiple loci that are involved in the pathogenesis of murine lupus.[2,3] One of the best characterized murine models of lupus is the New Zealand hybrid model, in which the F1 hybrid produced by crossing the New Zealand black (NZB) and New Zealand white (NZW) strains develops lupus-like disease. Genetic analyses by several groups have led to the identification of several distinct genomic segments that contain lupus susceptibility genes. In addition to MHC genes on chromosome 17, several other loci from NZB or NZW are associated with lupus-related phenotypes. Of these, regions on chromosomes 1, 4, and 7 have consistently

Mechanisms of Lymphocyte Activation and Immune Regulation VIII
Edited by Sudhir Gupta, Kluwer Academic/Plenum Publishers, 2001

been shown to be involved in determining susceptibility to SLE in multiple mouse models, suggesting that these intervals may contain genes or clusters of genes that strongly influence the development of lupus.

Our laboratory performed a genome scan of (NZM2410 × B6) F1 × NZM2410 backcross progeny and identified four genomic intervals containing GN susceptibility loci.[5] To determine the contributions of different intervals to lupus susceptibility, four interval-specific congenic strains, designated B6.NZM*Sle1*, B6.NZM*Sle2* and B6.NZM*Sle3*, were subsequently generated[12] by using marker-assisted selection protocols, each carrying a single NZM2410-derived susceptibility interval on a C57BL/6 background. These congenic strains were used to characterize the component phenotype contributed by each susceptibility locus. B6.NZM*Sle1* mice, carrying the *Sle1* interval on chromosome 1, develop high titers of spontaneous IgG anti-nuclear antibody (ANA), and show a strong spontaneous humoral response targeting primarily to H2A/H2B/DNA subnucleosomes, indicating that *Sle1* can cause the loss of self tolerance and mediate the development of T-dependent humoral autoimmunity to nuclear antigens.[12] Although *Sle1* was the strongest interval detected in the original cross, it alone in a C57BL/6 genome does not lead to significant levels of nephritis. *Sle1* also causes an expanded pool of histone-reactive T cells, indicating that *Sle1* causes a breach in tolerance to nuclear autoantigens in both the T and B cell compartments. B6.NZM*Sle2* mice, carrying *Sle2* on chromosome 4, spontaneously develop high levels of IgM antibodies against a variety of foreign and self antigens, indicating that this interval is responsible for intrinsic B-cell hyper-reactivity and elevated B1-cell formation. However, *Sle2* by itself on the normal B6 background is insufficient to generate IgG ANAs or nephritis, suggesting that *Sle2* may contribute to the pathogenesis of SLE by reducing the B cell signaling threshold. B6.NZM*Sle3* mice, carrying *Sle3* on chromosome 7, exhibit elevated serum levels of polyclonal/polyreactive IgM and IgG antibodies, with low-grade reactivity to all chromatin components.

2. EPISTATIC INTERACTIONS AMONG DISEASE SUSCEPTIBILITY GENES IN DIFFERENT BIOLOGICAL PATHWAYS LEAD TO THE DEVELOPMENT OF LUPUS

Genetic and immunologic analyses of the interval-specific congenic strains indicated that 1) each of these intervals contains a gene or genes that independently contribute to GN susceptibility; and 2) that several susceptibility intervals are needed to express highly penetrant GN. Although none of the aforementioned monocongenic strains develop lupus, Mohan *et al*[13] demonstrated that the combination of *Sle1* and *Sle3* on C57BL/6 background was sufficient to cause moderately penetrant fatal glomerulonephritis, indicating that epistatic interactions between these two intervals strongly impact the development and severity of this disease. The B6.NZM*Sle1/Sle3* mice also exhibit splenomegaly, significantly expanded populations of activated B and CD4 T cells, and a robust IgG antibody response targeting multiple components of chromatin, intact glomeruli, and basement membrane matrix antigens. The combination of *Sle1* and *Sle2* on C57BL/6 background has also been shown to lead to severe glomerulonephritis. More recently, Morel *et al*[14] demonstrated that the co-expression of *Sle1*, *Sle2*, *Sle3* as a B6-triple congenic results in severe systemic autoimmunity and fully penetrant, fatal glomerulonephritis, indicating that *Sle2* potentiates the development of fatal lupus. However, the combination of *Sle2* and *Sle3* on C57BL/6 background failed to mediate fatal disease. These results indicate that *Sle1*, through mediating the loss of tolerance to chromatin, is a key element in the development of lupus. Our analyses of these congenic strains support a working model of lupus pathogenesis in which different genes interact to mediate the development of lupus. In this model (Figure 1), we hypothesize that genes and/or environmental factors potentiate disease pathogenesis through their impact on three separate biological pathways:1) the loss of immunologic tolerance to nuclear antigens, 2) the dysregulation of the immune system, and 3) end-organ targeting. We further suggest that epistatic interactions between genes in these three pathways are an essential element in the development of severe lupus nephritis.

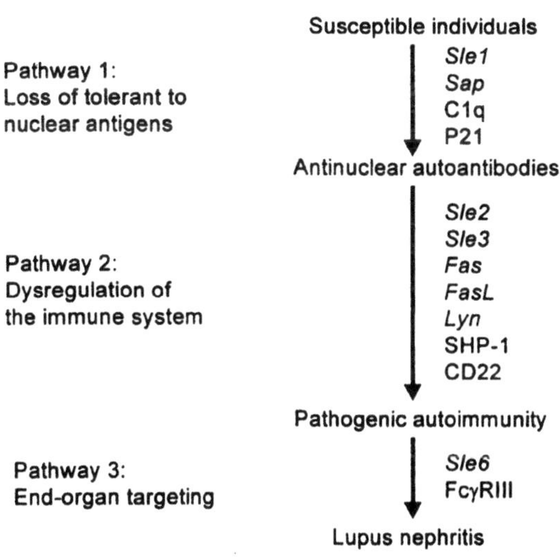

Figure 1. A working model of the development of murine lupus nephritis. In this model, three different biological pathways mediate the progressive development of severe disease pathogenesis.

As illustrated in Figure 1, the initial phase in the development of lupus pathogenesis is the loss of immune tolerance to nuclear autoantigens. In our murine model of lupus, *Sle1* mediates this disease phenotype. Although the precise identify of *Sle1* and the molecular mechanisms involved in lupus pathogenesis for this interval are unclear presently, the phenotype of the mono and poly congenic strains unambiguously place *Sle1* in the first biological pathway. We also postulate that two additional genes, *C1q* and *Sap*, belong to this pathway. Mice with targeted disruptions in the *C1q*[15] or *Sap*[16] gene spontaneously develop lupus-like disease. Since the C1q complement component and serum amyloid P component (Sap) have been shown to bind to and clear apoptotic cells and nuclear debris released by necrosis, it is believed that disruption of these genes leads to the accumulation of nuclear chromatin accessible for immune cells. Interestingly, a recent study by Balomenos *et al*[17] showed that cell cycle regulator p21-deficiency leads to the loss of tolerance to nuclear antigens and development of sex-linked lupus. Although p21-deficient mice develop normally, it is not surprising that the proliferative properties of lymphocytes render them vulnerable to the cell cycle dysregulation and susceptible to the development of autoimmune diseases. This observation can also place p21 gene in the first biological pathway.

Genes in the second biological pathway include *Sle2, Sle3, Fas, FasL, Lyn, SHP-1*, and CD22. Fas and FasL are important in the elimination of autoreactive immune cells, and mutations in these genes have been shown to accelerate the development of lupus nephritis.[18,19] The importance of CD22, *Lyn*, and *SHP-1* in regulating BCR signaling and selection has been verified by Cornall *et al.*[20] The role for the genes in the second pathway is mainly to remove autoreactive cells and/or downregulate the signaling process in immune cells. *Sle3*, another important element in the second pathway will be detailed below.

Genes placed in the third pathway are *Sle6* on chromosome 5 and *FcγRIII*. The impact of FcγRIII expression on lupus pathogenesis was demonstrated in Fc receptor gamma chain-deficient NZB/NZW mice, which generated and deposited immune complex and activated complement in kidney glomeruli, but failed to develop severe nephritis.[21] These results support the hypothesis that Fc receptor expression plays a key role in immune mediated the end organ damage during lupus development. Genes in this pathway could function by modifying immune inflammatory processes and/or modifying the end organ itself. Although none of the genes mentioned above are absolutely required in the pathogenesis of lupus nephritis, the epistatic interactions among these three separate pathways are crucial for the development of severe disease.

3. *SLE3* MEDIATES THE DYSREGULATION OF T CELL ACTIVITY

Accumulating results from our studies strongly support the placement of *Sle3* in the second biological pathway in lupus pathogenesis. Characterization of *Sle3*[22] indicated that this interval mainly affects T lymphocytes, causing a generalized T-cell activation. Compared to C57BL/6 mice, B6.NZM*Sle3* mice exhibit elevated CD4:CD8 ratios in spleens and lymph nodes, an age-dependent accumulation of activated CD4+ T-cells, a more diffuse splenic architecture, and a stronger immune response to T-dependent antigens. In vitro studies also indicated that *Sle3*-bearing T cells have stronger proliferation, increased expansion of CD4+ T cells, and reduced apoptosis, following stimulation with anti-CD3.

The original genomic scan of (C57BL/6J x NZM2410) x NZM2410 identified two intervals within *Sle3* on chromosome 7 that are linked with lupus development: one is close to the centromere, while the other is close to the *pink-eyed dilution (p)* locus, about 30 cM away. Recent studies with bi-congenic mice, B6.NZM*Sle1/Sle3*, and tri-congenic mice, B6.NZM*Sle1/Sle2/Sle3*, showed that the whole *Sle3* interval (from the centromere through marker *D7Mit31*) strongly potentiates the development of fatal lupus.[13,14] Since *Sle1* breaks immune tolerance to nuclear chromatin but fails to mediate severe disease when isolated on a B6 background, these observations indicate that *Sle3* interval is responsible for providing a crucial dysregulation in T cell help and/or weakening the inhibitory signaling pathway for B cell activation. At present, several genes that are related to immune activation have been mapped to this interval. Among these are TGF-β, CD22, CD79α, Bax, Bcl3, etc. Although CD22 in NZM2410 has allelic differences with CD22 in the B6 genome,[23] and this molecule is important in downregulating B cell activation, the impact of this polymorphism in lupus pathogenesis remains to be elucidated. It has been shown that TGFβ1 is an important negative regulator of T cell homeostasis,[24] but so far, no aberrant expression of this gene has been identified in NZM2410. Undoubtedly, our ongoing analysis of positional candidates for *Sle3*, together with the efforts from other groups on this interval, will our understanding of the molecular mechanisms by which *Sle3* accelerates lupus pathogenesis.

4. REFERENCES

1. Kotzin, BL: Systemic lupus erythematosus. Cell 85:303-306, 1996
2. Wakeland, EK, Wandstrat, AE, Liu, K, and Morel, L: Genetic dissection of systemic lupus erythematosus. Curr. Opin. Immunol. 11:701-707, 1999
3. Vyse TJ, Kotzin BL: Genetic susceptibility to systemic lupus erythematosus. Ann. Rev. Immunol. 16:261-292,1998

4. Foster MH: Relevance of systemic lupus erythematosus nephritis animal models to human disease. Semin. Nephrol. 19:12-24,1999

5. Morel L, Rudofsky UH, Longmate JA, Schiffenbauer J, Wakeland EK: Polygenic control of susceptibility to murine systemic lupus erythematosus. Immunity 1:219-229, 1994

6. Drake CG, Babcock SK, Palmer E, Kotzin BL: Genetic analysis of the NZB contribution to lupus-like autoimmune disease in (NZB x NZW)F1 mice. Proc. Natl. Acad. Sci. U.S.A. 91:4062-4066, 1994

7. Kono DH, Burlingame RW, Owens DG, Kuramochi A, Balderas RS, Balomenos D, Theofilopoulos AN: Lupus susceptibility loci in New Zealand mice. Proc. Natl. Acad. Sci. U. S. A. 91:10168-10172,1994

8. Hogarth MB, Slingsby JH, Allen PJ, Thompson EM, Chandler P, Davies KA, Simpson E, Morley BJ, Walport MJ: Multiple lupus susceptibility loci map to chromosome 1 in BXSB mice. J. Immunol. 161:2753-2761, 1998

9. Watson ML, Rao JK, Gilkeson GS, Ruiz P, Eicher EM, Pisetsky DS, Matsuzawa A, Rochelle JM, Seldin MF: Genetic analysis of MRL-lpr mice: relationship of the Fas apoptosis gene to disease manifestations and renal disease-modifying loci. J. Exp. Med. 176:1645-1656, 1992

10. Vidal S, Kono DH, Theofilopoulos AN: Loci predisposing to autoimmunity in MRL-Fas ^{lpr} and C57BL/6-Fas^{lpr} mice. J. Clin. Invest.101:696-702, 1998

11. Morel L, Mohan C, Yu Y, Schiffenbauer J, Rudofsky UH, Tian N, Longmate JA, Wakeland EK: Multiplex inheritance of component phenotypes in a murine model of lupus. Mamm, Genome, 10:176-181, 1999

12. Morel L, Yu Y, Blenman KR, Caldwell RA, Wakeland EK: Production of congenic mouse strains carrying genomic intervals containing SLE-susceptibility genes derived from the SLE-prone NZM2410 strain. Mamm. Genome 7:335-339, 1996

13. Mohan C, Morel L, Yang P, Watanabe H, Croker B, Gilkeson G, Wakeland EK: Genetic dissection of lupus pathogenesis: a recipe for nephrophilic autoantibodies. J. Clin. Invest. 103:1685-1695, 1999

14. Morel L, Croker BP, Blenman KR, Mohan C, Huang G, Gilkeson G, Wakeland EK: Genetic reconstitution of systemic lupus erythematosus immunopathology with poly-congenic murine strains. Proc. Natl. Acad. Sci. U.S.A. 97:6670-6675, 2000

15. Botto M, Dell'Agnola C, Bygrave AE, Thompson EM, Cook HT, Petry F, Loos M, Pandolfi PP, Walport MJ: Homozygous C1q deficiency causes glomerulonephritis associated with multiple apoptotic bodies. Nat. Genet. 19:56-59, 1998

16. Bickerstaff MC, Botto M, Hutchinson WL, Herbert J, Tennent GA, Bybee A, Mitchell DA, Cook HT, Butler PJ, Walport MJ, Pepys MB: Serum amyloid P component controls chromatin degradation and prevents antinuclear autoimmunity. Nat. Med. 5:694-697, 1999

17. Balomenos D, Martin-Caballero J, Garcia MI, Prieto I, Flores JM, Serrano M, Martinez AC: The cell cycle inhibitor p21 controls T-cell proliferation and sex-linked lupus development. Nat. Med. 6:171-176, 2000

18. Watanabe-Fukunaga R, Brannan CI, Copeland NG, Jenkins NA, Nagata, S: Lymphoproliferation disorder in mice explained by defects in Fas antigen that mediates apoptosis. Nature 356:314-317, 1992

19. Roths JB, Murphy ED, Eicher EM: A new mutation, gld, that produces lymphoproliferation and autoimmunity in C3H/HeJ mice. J. Exp. Med. 159:1-20,1984

20. Cornall RJ, Cyster JG, Hibbs ML, Dunn AR, Otibopy KL, Clark EA, Goodnow CC: Polygenic autoimmune traits: Lyn, CD22, and SHP-1 are limiting elements of a biochemical pathway regulating BCR signaling and selection. Immunity 8:497-508, 1998

21. Clynes R, Dumitru C, Ravetch JV: Uncoupling of immune complex formation and kidney damage in autoimmune glomerulonephritis. Science 279:1052-1054, 1998

22. Mohan C, Yu Y, Morel L, Yang P, Wakeland EK: Genetic dissection of SLE pathogenesis: *Sle3* on murine chromosome 7 impacts T cell activation, differentiation, and cell death. J. Immunol. 162:6492-6502, 1999
23. Lajaunias F, Ibnou-Zekri N, Fossati-Jimack L, Chicheportiche Y, Parkhouse RM, Mary C, Reininger L, Brighouse G, Izui S: Polymorphisms in the Cd22 gene of inbred mouse strains. Immunogenetics 49:991-995, 1999
24. Gorelik L, Flavel RA: Abrogation of TGF signaling in T cells leads to spontaneous T cell differentiation and autoimmune disease. Immunity 12:171–181, 2000

FACTORS CONTRIBUTING TO AUTOIMMUNE DISEASE

Kristine M. Garza, Linh T. Nguyen, Russell G. Jones, and Pamela S. Ohashi

Ontario Cancer Institute
Departments of Medical Biophysics and Immunology
610 University Ave.
Toronto ON   M5G 2M9
Canada

INTRODUCTION

T cells that emigrate from the thymus with specificities towards self-antigens face one of two fates.   They either remain in the peripheral repertoire if they are immunologically unaware or ignorant of their cognate ligand., or if they encounter sufficient concentrations of their cognate ligand, the self-reactive T cells are subject to peripheral tolerance.   Pathological autoimmunity can arise when either or both of these mechanisms are breached.

Autoimmunity may be induced through the activation of 'ignorant' autoreactive T cells due to events promoted by inflammation.   Such events include the acquisition of sufficient concentrations of stimulating antigen, the presentation of antigen on cells with costimulatory abilities, the migration of ignorant T cells to areas of antigen presentation, and/or the migration of antigen-laden APCs to T cell areas of lymphoid organs.   This is a multi-component process, requiring the coordination of a number of events to provide a sufficiently activating environment and is highly influenced by a number of factors.

Inhibiting the induction of peripheral T cell tolerance, deletion or anergy, can also induce pathogenic autoimmunity.   The effect of preventing deletion of autoreactive T cells is highlighted in patients with Autoimmune Lymphoproliferative Syndrome (ALPS) [1].   In ALPS patients, T cell receptor (TCR)-induced apoptosis is defective and is associated with non-malignant lymphoproliferation and autoimmunity.   APLS is an autosomal dominant disorder, however, not all carriers become diseased.   The implication is that other parameters affect this form of autoimmune disease induction as well.

Several candidate factors which affect the induction of autoimmunity have been elucidated. Studies have demonstrated that autoimmune disease is more prevalent in people with certain MHC types [2,3]. The association of autoimmune disease with a particular MHC allele is thought to be due to the ability of the MHC to bind and present target self antigens in addition to the ability of MHC to select for self reactive T cells. However, not all individuals who carry disease-susceptible MHC alleles develop disease.   Environmental agents have also been proposed to influence the induction of autoimmunity via a number of mechanisms including molecular mimicry [4,5], the induction of tissue damage that in turn causes the release of cryptic self peptide antigens [6,7], or the upregulation or ectopic expression of costimulatory molecules [8,9], all which have also been shown to result in self-directed immune responses. The induction of autoimmunity is also highly influenced by the presence of cytokines [10,11]. Both pro- and anti-inflammatory cytokines can promote or prevent the induction of disease depending on the timing, level, and location of production. Lastly, regulatory lymphocytes may also influence the induction of autoimmune disease. The presence of regulatory lymphocytes has been demonstrated in a number of experimental systems in which the T cell repertoire is restricted, resulting in the spontaneous induction of autoimmune disease suggesting that the missing lymphocyte population has the responsibility for keeping the autoreactive T cells in check [12-14]. Thus, the loss of regulatory cells may also promote autoimmunity.

Not one factor, however, has been clearly identified as the key component to the induction of autoimmunity. It is most likely that different combination of factors leads to pathogenic disease. To further investigate factors that can influence the induction of autoimmune disease, we have focused on what affects the activation or the deletion of lymphocytic choriomeningitis virus (LCMV)-specific CD8$^+$ T cells. We have utilized transgenic mice that express the LCMV glycoprotein (LCMV-gp) on pancreatic β-islet cells, under the control of the rat insulin promoter (RIP-gp), as well transgenic mice that express the LCMV-gp-specific T cell receptor (P14) [15].

ROLE OF ACTIVATED APC IN PROMOTING AUTOIMMUNITY

A number of studies have convincingly demonstrated that antigen presenting cells (APC) can modulate the outcome of T cell receptor interaction with self-antigen. Models have proposed that this is dependent on costimulation and the presence of second signals [16]. Alternate models suggest that different subpopulations of APC define the outcome of T cell interaction with ligand, where one population induces tolerance and a second induces T cell activation [17-19]. Recent studies have suggested that the 'activation status' of the APC may dictate the outcome of T cell responses. The presentation of antigen on resting B cells to naïve T cells has been shown to induce T cell tolerance [20,21]. In contrast, the presentation of antigen in a pro-inflammatory environment has been shown to induce immunity or autoimmunity [22,23].

To assess the role of activated APC in the induction of autoimmunity, we treated RIP-gp/P14 double transgenic mice with the LCMV-derived peptide, p33, and the anti-CD40 activating antibody, FGK45 [24]. Recent reports have demonstrated that the maturation and activation of APCs can be induced with the ligation of CD40, resulting in increased capacity to present antigen [25], and the induction of CD8$^+$ immunity [26-28]. In addition, studies have shown that administration of FGK45, a rat anti-mouse CD40 activating antibody, *in vivo* leads to the activation of APCs and the induction of T cell function [27,28].

In untreated RIP-gp/P14 double transgenic mice, the LCMV-gp specific CD8$^+$ T cells are not tolerized, but remain immunologically unaware of the presence of their antigen and are fully capable of responding to antigen under the appropriate conditions and can induce insulitis and diabetes [15]. When treated with p33 and anti-CD40 antibody, all the transgenic mice became diabetic (Table 1) [29]. Transgenic mice that received p33 and the isotype matched control antibody (rat polyclonal anti-serum) demonstrated that peptide alone could not induce disease. In addition, mice that received an irrelevant control peptide (AV) and anti-CD40 antibody did not become diabetic. Thus, peptide administration and the *in vivo* activation of APC was critical for the induction of autoimmunity.

The status of T cell activity in the spleens of animals given peptide and anti-CD40 or control antibody was also assessed. T cell activity was measured by the upregulation of T cell activation markers, the induction of effector function, and the infiltration of the pancreas. The induction of activation markers, as well as cytotoxic activity, was identical in both groups (data not shown). Treatment with peptide and control antibody induced mild pancreatic infiltration (Table 1). In contrast, the combination of peptide and anti-CD40 antibody induced severe insulitis (Table 1).

Activation of APCs via CD40 has been shown to lead to increased production of IL-12, which promotes the release of IFNγ [30,31]. Treatment of double transgenic mice with peptide and control antibody did not promote the production of measurable levels of circulating IFNγ. However, the addition of anti-CD40 to the peptide treatment induced levels of IFNγ that were detectable in the serum (Table 1). Moreover, serum IFNγ levels correlated with increased expression of class I in the pancreatic islets (data not shown). Thus, anti-CD40 treatment led to the activation of APC's *in vivo*, with an enhanced production of IFNγ and enhanced pancreatic islet expression of class I MHC. This may have contributed to enhanced CTL infiltration of the pancreas and the onset of diabetes.

Table 1. Induction of diabetes, insulitis, and IFNγ production in RIP-gp/P14 mice treated with peptide and anti-CD40[1].

GROUP	DIABETES[2]		INSULITIS[3]		IFNγ[5]
	Incidence (%)	Mean onset (day)	Incidence (%)	Mean Severity[4]	(pg/ml)
In RIP-gp/P14 mice:					
p33 + anti-CD40	100 (20/20)	7	83	3.0	250 ± 40
p33 + control Ab	0 (0/20)	-	22	1.7	-
AV + anti-CD40	0 (0/10)	-	-	-	-
In RIP-gp/P14/CD28-deficient mice:					
p33 + anti-CD40	100 (8/8)	7	-	-	-

[1] Mice were treated intravenously with 5 μg of peptide on day 0 and 2, and with 100 ug of antibody on day 2.
[2] Blood glucose levels were monitored; diabetes = > 14 mm/L.
[3] Pancreatic islet infiltration by CD8$^+$ cells was determined on day 3. Three non-serial sections per mouse were assessed for islet number and severity of infiltration (n = 5 animals/group, 10-20 islets/mouse).
[4] Mean severity of only infiltrated islets, where a severity of 1 = peri-insulitis/mild insulitis, 2 = partial insulitis, and 3 = complete insulitis.
[5] Serum IFNγ levels were determined on day 3 by ELISA; detection limit = 65 pg/ml.

CD40 mediated APC activation has also been shown to upregulate the costimulatory molecules B7-1 and B7-2 and thereby also enhances T cell function [25,32]. To determine whether the induction of autoimmunity, via APC activation, was dependent upon engagement of costimulatory ligands, the P14 and RIP-gp transgenes were bred onto a CD28 deficient background. Immunization of RIP-gp/P14/CD28-negative mice with p33 and anti-CD40 resulted in the induction of diabetes with the same incidence and kinetics as seen in CD28 competent mice (Table 1). Therefore, CD28/B7 interactions were not crucial for the induction of CD8 mediated autoimmunity in this model.

In addition to their ability to promote immunity, bacterial or viral infections have been reported to inhibit the induction of peripheral tolerance as well [22,23]. Therefore, to examine whether the activation of APC can influence T cell tolerance, P14 TCR transgenic mice were treated intravenously with p33, to induce deletion of the transgenic T cells, and were given anti-CD40 or the isotype-matched control antiserum. As shown in Table 2, treatment with the activating anti-CD40 antibody dramatically inhibited the specific deletion of the LMCV-specific CD8$^+$Vα2$^+$ transgenic T cells. Thus, activated APC not only promote the induction of autoimmunity, but can prevent the induction of tolerance as well.

Table 2. Treatment of P14 transgenic mice with anti-CD40 prevents p33-induced T cell tolerance[1].

GROUP	NUMBER of CD8+/Vα2+ CELLS (x 10^6)[2]				
	Day 0	Day 3	Day 6	Day 9	Day 12
p33 + anti-CD40	20.1 ± 5.40	21.5 ± 0.85	43.7 ± 7.30	67.5 ± 5.50	132.7 ± 24.13
p33 + control Ab	19.2 ± 3.60	33.6 ± 4.0	4.0 ± 0.30	2.27 ± 2.30	3.14 ± 0.15

[1]P14 transgenic mice were treated 3 times, 3 days apart, starting on d0, with 5 µg of p33 intravenously. 2d after each peptide treatment, mice were given 100 µg of antibody.

[2]Splenic numbers of transgenic T cells were determined in different cohorts of mice over time by flow cytometry (n = 2-3 mice/group/time point).

This study provides evidence to confirm that T cell interactions with peptide presented in a "resting" or naïve environment is not sufficient to lead to autoimmunity. The activation of APCs via anti-CD40, however, leads to enhanced cytokine production, contributing to a proinflammatory environment and autoimmune diabetes. Moreover, activated APC can prevent the induction of T cell tolerance. Thus, a factor leading or predisposing an individual to autoimmune disease may be the chronic activation of APCs. The recognition of self-antigen by autoreactive T cells in the presence of chronically activated APC could thus bias T cell interactions from ignorance or tolerance to autoimmunity (Fig. 1).

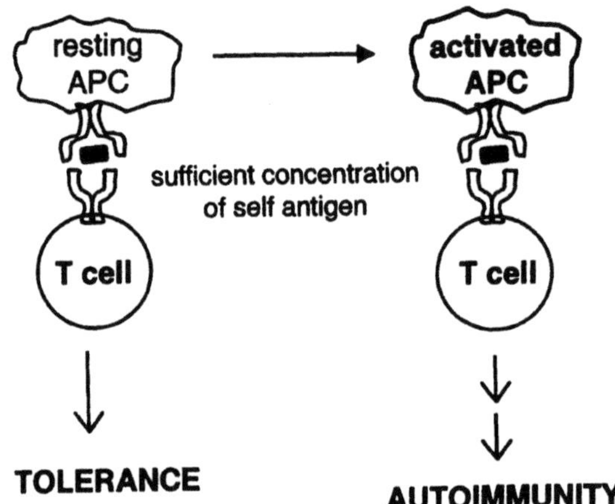

Figure 1. APCs presenting self antigen in a 'resting' or naïve environment are likely to promote the induction of tolerance. However, in a pro-inflammatory environment, which activates APCs, the presentation of self antigen may lead to the inhibition of tolerance induction and may promote the induction of autoimmunity.

ROLE OF DEATH RECEPTORS, CD95 AND TNFR1, IN PERIPHERAL T CELL TOLERANCE

Antigen-induced apoptosis of peripheral T cells plays an important role in the elimination of autoreactive T cells [33]. Of the many cell surface receptors demonstrated to be involved in T cell apoptosis, CD95 is of particular importance as highlighted by the lymphoproliferation mutant mouse strain (*lpr*) (CD95-deficient) which is characterized by lymphadenopathy, splenomegaly, accumulation of $CD4^-CD8^-B220^-$ T cells and the production of various autoantibodies [34,35]. In addition, studies using superantigens, soluble peptides, or cross-presented self-antigens in *lpr* mice has also demonstrated a role for CD95 in peripheral T cell apoptosis [36-39]. TNFR1 has also been implicated in T cell apoptosis. Elimination of TNFR1 in *lpr* mice accelerates the onset of the *lpr* phenotype [40], and further evidence has suggested that TNF is involved in peripheral T cell deletion. [41-43].

To assess the potential cooperation between CD95 and TNFR1 in mediating peripheral T cell tolerance through deletion, mice deficient in both receptors (*tnfr1$^{-/-}$lpr/lpr* mice) were generated [44]. In addition, these mice were also crossed with the LCMV-specific TCR transgenic mice, P14. Peptide-induced peripheral tolerance can be induced in P14 mice with the administration of 500 µg of p33 emulsified in IFA given every 3 days intraperitoneally. This protocol is thought to mirror prolonged or chronic exposure to peripherally expressed self-antigens. The induction of tolerance is marked by a transient expansion followed by a rapid deletion of LCMV-transgenic T cells. P14 mice deficient in both TNFR1 and CD95 displayed normal deletion kinetics of the transgenic T cells upon treatment with peptide (data not shown). The induction of tolerance was confirmed by the inability of splenocytes from treated animals to proliferate in vitro to p33 seventeen days after the initial peptide treatment.

T cell deletion was also assessed after the administration of a pathogenic stimulus in vivo using LCMV by determining the kinetics of CTL responses. There was no difference in the decline of CTL activity between any of the mice (data not shown). Maximal CTL activity was observed 8 days after infection followed by a marked decline by day 13, regardless of genotype. In addition, the kinetics of p33-specific T cell expansion and deletion was assessed in recipient mice of P14 T cells infected with LCMV. The total number of TCR-transgenic T cells in the spleen was similar at 1 and 7 weeks after infection (data not shown). Thus, in the absence of both TNFR1 and CD95, deletion following viral challenge proceeds normally.

Peptide-induced tolerance in the absence of TNFR1 and CD95 was also evaluated in non-TCR transgenic mice, which bear a more physiological frequency of peptide-specific T cells in contrast to P14 mice. Treatment of mice, heterozygous for TNFR1 and CD95, three times with 500 µg p33 emulsified in IFA induced tolerance of p33-specific T cells, as measured by the lack of p33-specific cytotoxic activity in response to subsequent LCMV challenge (Fig. 2A, *tnfr1$^{+/-}$lpr/+*). In contrast, tolerance was not induced in peptide-treated TNFR1/CD95 deficient mice as determined by the induction of p33-specific CTL responses upon LCMV infection (Fig. 2A, *tnfr1$^{-/-}$lpr/lpr*). Mice deficient in TNFR1 alone were partially tolerized (Fig. 2B, *tnfr1$^{-/-}$fas$^{+/+}$*), indicating that TNFR1 contributes to $CD8^+$ T cell deletion. Mice deficient in CD95 alone showed normal tolerance induction, suggesting that TNFR1 or other molecules can compensate for deletion in the absence of CD95 (Fig. 2B, *tnfr1$^{+/+}$ lpr/lpr*). However, it is evident that CD95 does contribute to $CD8^+$ deletion being that TNFR1/CD95 double knockout mice have a more severe defect in deletion compared to TNFR1 single knockout mice. Therefore, these data demonstrate that TNFR1 and CD95 cooperate to mediate T cell deletion induced by high dose peptide in IFA.

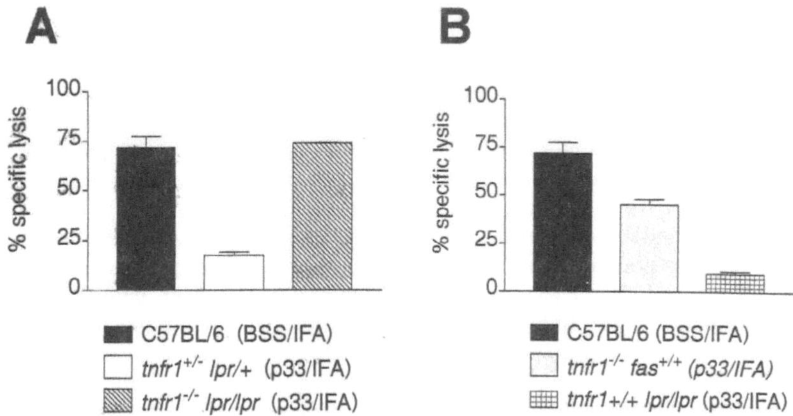

Figure 2. Peptide-induced tolerance is defective in *tnfr1-/- lpr/lpr* mice. The indicated mice were immunized i.p. with 500 μg of p33 (or BSS as control) emulsified in IFA on days 0, 3, and 6. Ten days after the last peptide treatment, mice were infected with 2000 PFU LCMV. Eight days after infection, splenocytes were assayed in 5-h chromium release assay using EL-4 target cells pre-pulsed with ^{51}Cr and p33. Lysis of targets pulsed with ^{51}Cr alone was <15%. The mean of duplicate wells from representative mice is shown at an effector to target ratio of 23:1.

Together, the data suggest that T cell deletion can occur dependently or independently of the death receptors, CD95 and TNFR1. Where antigen is limited, such as when virus is efficiently cleared, CD95/TNFR1-independence is consistent with the model that deletion occurs due to withdrawal of lymphokines and not to the presence of death receptors [1,45-47]. When antigen exposure is prolonged and occurs in the presence of physiological numbers of antigen-specific T cells, T cell deletion is dependent upon CD95 and TNFR1. This suggests that in the absence of this cooperative effort, autoreactive CD8+ T cells may not be appropriately deleted, predisposing an individual to autoimmune disease (Fig. 3).

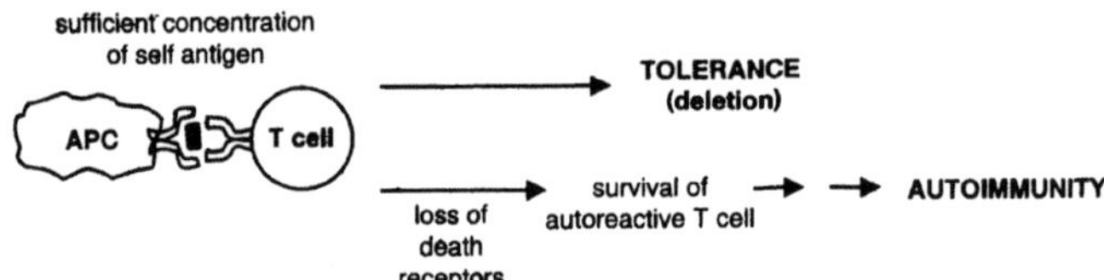

Figure 3. Recognition of sufficient levels of self-antigen by autoreactive T cells in a non-inflammatory environment promotes the induction of deletion (tolerance) mediated through the death receptors CD95 and TNFR1. Loss of these receptors may promote the induction of autoimmunity by inhibiting the removal of autoreactive T cells.

ROLE OF SURVIVAL MOLECULES

For autoimmune disease to manifest, not only do the activated autoreactive T cells have to remain alive for a sufficient amount of time to home to the target organ and perform effector functions, sufficient numbers of autoreactive T cells must also exist to cause significant and clinically detectable tissue damage. This may explain why autoimmune disease occurs so readily in many TCR transgenic models. Thus, another potential factor influencing the induction of autoimmunity may be how long an autoreactive T cell survives after recognition of self-antigen.

The lifespan of a T cell upon TCR ligation is dependent on a balance between survival and apoptosis. The induction of apoptosis in mature T cells is strictly regulated after antigenic triggering [1], however, costimulatory molecules, such as CD28, are believed to contribute to survival signals and promote immune responses [48-50]. Recent studies have suggested a role for phosphatidylinositol 3-kinase (PI3K), a downstream target of TCR and CD28-mediated signaling, in the suppression of apoptosis [48]. The serine/threonine kinase Protein Kinase Bβ (PKB/Akt), a downstream target of PI3K activity, is also emerging as a key molecule involved in regulating cell survival in a variety of models [51]. Activated PKB has been demonstrated to play a role in growth factor- and cytokine-mediated survival, and protects cells from apoptosis induced by a variety of stimuli [52].

To determine if PKB was a relevant target in T cell signaling, the activation of PKB was assessed in mature peripheral T cells upon ligation of the TCR [53]. Activation of peripheral T cells with anti-CD3 alone or anti-CD3 plus anti-CD28 led to a substantial increase in PKB activation relative to untreated cells (data not shown). In addition, the failure of these stimulatory antibodies to activate PKB in cells pretreated with the PI3K inhibitor wortmannin [54] indicated that TCR-mediated activation of PKB was dependent upon PI3K activity. These results established that PKB represents a physiologically relevant target downstream of the TCR in primary T cells.

The role of PKB in T cell survival was assessed by generating mice that expressed a constitutively form of PKB (gag-PKB) only in the T cell lineage [53]. Purified T cells from gag-PKB and non-transgenic littermates were assessed for the expression of the anti-apoptotic proteins Bcl-2 and Bcl-X$_L$. The expression of these two molecules is tightly regulated in mature T cells. Bcl-2 expression is high in mature peripheral T cells and remains high following activation [55], while Bcl-X$_L$ is upregulated only upon activation and has been proposed to confer transient survival to activated T cells [48]. Endogenous levels of Bcl-X$_L$ were increased in transgenic mice relative to negative littermate control mice (data not shown). There was no difference in Bcl-2 protein levels in the T cells of these mice (data not shown), suggesting that PKB does not affect Bcl-2 expression in mature T cells. These results indicated that PKB influences the level of Bcl-X$_L$ protein in mature T cells and suggested that T cells from PKB transgenic mice may have enhanced survival capabilities.

To investigate the functional consequences of PKB activation downstream of the TCR, the PKB mice were crossed to P14 single transgenic mice. Splenocytes from P14 TCR transgenic and P14/gag-PKB double transgenic animals were cultured in the presence of APCs pre-incubated with p33 peptide, and proliferation was assessed over time. While levels of proliferation remained similar in both groups up to 48 hours, T cells expressing the gag-PKB transgene displayed enhanced proliferation relative to P14 control lymphocytes up to 120 hours post-activation (Table 3). Proliferation against an irrelevant control peptide was less than 4% of maximal proliferation for all time points, regardless of gag-PKB transgene expression (data not shown). These data suggested that PKB activity can alter antigen-specific T cell proliferation.

Table 3. PKB enhances antigen-specific proliferative responses of mature T cells[1].

GROUP	% of MAXIMAL PROLIFERATION[2]			
	48 h	72 h	96 h	120 h
P14	100.0 ± 5.30	75.0 ± 12.9	29.0 ± 9.6	2.0 ± 0.87
P14/gag-PKB	99.0 ± 1.5	96.0 ± 10.0	55.0 ± 9.7	12.0 ± 4.3

[1]Splenocytes from mice were cultured in the presence of APCs pre-pulsed with 10^{-5} M p33 and later pulsed with [^{3}H]-thymidine at the indicated time points.
[2]To standardize for different experiments, scintillation counts obtained were normalized and expressed as a percentage of maximal proliferation (max=100%).

Studies are underway to determine if T cells from P14/gag-PKB mice can influence the induction of autoimmune disease. A recent study utilizing mice heterologous for a PTEN deficiency has suggested that survival molecules can promote autoimmunity. PTEN is a tumor-suppressing gene encoding a phosphatase. A substrate for PTEN is phosphatidylinositol triphosphate (PIP-3), a lipid second messenger produced by PI3K. In the absence of *Pten* activity, PIP-3 concentrations increase, leading to enhanced phosphorylation and activation of PKB [56]. Mice heterozygous for an inactive PTEN mutation, presented with an autoimmune disorder [57]. Splenocytes from these mice were marked by the presence of constitutively activated PKB and enhanced proliferative T cell responses. Autoimmunity was attributed to the inability to delete autoreactive lymphocytes because of defective CD95-mediated killing. Crosses between Bcl-2 overexpressing trangenic mice and *lpr* mice have also suggested a role for survival molecules in autoimmunity in which a synergistic increase in lymphadenopathy was seen [58,59]. Thus, the effect on autoimmune disease induction by survival molecules may be two-fold: inhibition of normal peripheral deletion processes and enhanced survival/proliferative responses upon antigen recognition (Fig. 4).

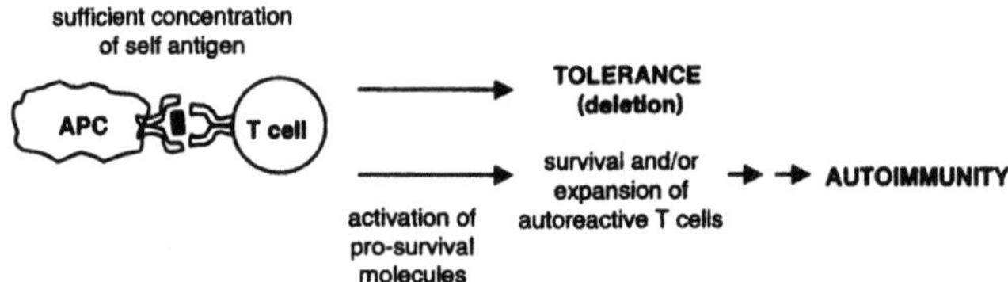

Figure 4. Recognition of sufficient levels of self-antigen by autoreactive T cells in a non-inflammatory environment promotes the induction of deletion (tolerance). Upregulation of pro-survival molecules may inhibit deletion and may promote enhanced survival and perhaps expansion of autoreactive T cells upon recognition of self antigen, thus promoting the induction of autoimmunity.

CLOSING REMARKS

It is evident that the induction of autoimmunity is not a simple process but rather the result of a series of events that promote the survival and activation of autoreactive T cells. Herein, we have identified activated APC, the loss of death receptors, and the acquisition of pro-survival molecules as factors that may inhibit the induction of tolerance and promote the activation of autoreactive T cells. Each alone, however, is insufficient for the induction of autoimmunity, this requires the coordination of a number of events to productively induce pathogenic T cell responses. The key direction for future autoimmune disease research endeavors is to further understand the requirements for T cell activation, survival, and death as well as the requirements for the maintenance of an inflammatory immune response which promote T cell responses.

REFERENCES

1. Lenardo MJ, Chan FKM, Hornung F, et al: Mature T Lymphocyte Apoptosis -immune regulation in a dynamic and unpredictable environment. Annu Rev Immunol 17:221-253, 1999

2. Nepom GT, Erlich H: MHC class II molecules and autoimmunity. Annu Rev Immunol 9:493-525, 1991

3. Wicker LS: Major histocompatability complex-linked control of autoimmunity. J Exp Med 186:973-975, 1997

4. Ohteki T, Hessel A, Bachmann MF, et al: Identification of a cross-reactive self ligand in virus-mediated autoimmunity. Eur J Immunol 29:2886-2896, 1999

5. Bachmaier K, Neu N, de la Maza LM, Pal S, Hessel A, Penninger J: Chlamydia infections and heart disease linked through antigenic mimicry. Science 26:1238-1239, 1999

6. Gammon G, Sercarz EE, Benichou G: The dominant self and the cryptic self: shaping the autoreactive T cell repertoire. Immunol Today 12:193-195, 1991

7. Lehmann PV, Sercarz EE, Forsthuber T, Dayan CM, Gammon G: Determinant spreading and the dynamics of the autoimmune T cell repertoire. Immunol Today 145:203-208, 1993

8. von Herrath MG, Guerder S, Lewicki H, Flavell RA, Oldstone MBA: Coexpression of B7-1 and viral ("self") transgenes in pancreatic β cells can break peripheral ignorance and lead to spontaneous autoimmune diabetes. Immunity 3:727-738, 1995

9. Harlan DM, Hengartner H, Huang ML, et al: Mice expressing both B7 and viral glycoprotein on pancreatic beta cells along with glycoprotein-specific transgenic T cell develop diabetes due to a breakdown of T lymphocyte unresponsiveness. Proc Natl Acad Sci USA 91:3137-3141, 1994

10. Falcone M, Sarvetnick N: Cytokines that regulate autoimmune responses. Curr Opin Immunol 11:670-676, 1999

11.   Singh VK, Mehrotra S, Agarwal SS: The paradigm of Th1 and Th2 cytokines: its relevance to autoimmunity and allergy. Immunol Res 20:147-161, 1999

12.   Gleeson PA, Toh B-H, van Driel IR: Organ-specific autoimmunity induced by lymphopenia. Immunol Rev 149:97-125, 1996

13.   Mason D, Powrie F: Control of immune pathology by regulatory T cells. Curr Opin Immunol 10:649-655, 1998

14.   Seddon B, Mason D: The third function of the thymus. Immunol Today 21:95-99, 2000

15.   Ohashi PS, Oehen S, Bürki K, et al: Ablation of "tolerance" and induction of diabetes by virus infection in viral antigen transgenic mice. Cell 65:305-317, 1991

16.   Gill RG, Coulombe M, Lafferty KJ: Pancreatic islet allograft immunity and tolerance: the two-signal hypothesis revisited. Immunol Rev 149:75-96, 1996

17.   Shortman K, Wu L, Süss G, et al: Dendritic cells and T lymphocytes: developmental and functional interactions. Ciba Foundation Symposium 204:130-141, 1997

18.   Fazekas de St.Groth B: The evolution of self-tolerance: a new cell arises to meet the challenge of self-reactivity. Immunol Today 19:448-454, 1998

19.   Banchereau J, Steinman RM: Dendritic cells and the control of immunity. Nature 392:245-252, 1998

20.   Eynon EE, Parker DC: Small B cells as antigen-presenting cells in the induction of tolerance to soluble protein antigens. J Exp Med 175:131-138, 1992

21.   Fuchs E, Matzinger PB: cells turn off virgin butn ot memory T cells. Science 258:1156-1159, 1992

22.   Rocken M, Urban JF, Shevach EM: Infection breaks T-cell tolerance. Nature 359:79-82, 1992

23.   Ehl S, Hombach J, Aichele P, et al: Viral and bacterial infections interfere with peripheral tolerance induction and activate CD8[+] T cells to cause immunopathology. J Exp Med 187:763-774, 1998

24.   Rolink A, Melchers F, Andersson B: The SCID but not the RAG-2 gene product is required for S mu-S epsilon heavy chain class switching. Immunity 5:319-330, 1996

25.   Grewal IS, Flavell RA: CD40 and CD154 in cell-mediated immunity. Annu Rev Immunol 16:111-135, 1998

26.   Ridge JP, Di Rosa F, Matzinger P: A conditioned dendritic cell can be a temporal bridge between a CD4[+] T-helper and a T-killer cell. Nature 393:474-478, 1998

27.   Bennet SRM, Carbone FR, Karamalis F, Flavell RA, Miller JFAP, Heath WR: Help for cytotoxic-T-cell responses is mediated by CD40 signalling. Nature 393:478-480, 1998

28.   Schoenberger SP, Toes REM, van der Voort EIH, Offringa R, Melief CJM: T-cell help for cytotoxic T lymphocytes is mediated by CD40-CD40L interactions. Nature 393:480-483, 1998

29.   Garza KM, Chan SM, Suri R, et al: Role of antigen presenting cells in mediating tolerance and autoimmunity. J Exp Med 191:2021-2027, 2000

30.   Cella M, Sheidegger D, Palmer-Lehmann K, Lane P, Lanzavecchia A, Alber G: Ligation of CD40 on dendritic cells triggers production of high levels of interleukin-12 and enhances T cell stimulatory capacity: T-T help via APC activation. J Exp Med 184:747, 1996

31.   Ohteki T, Fukao T, Suzue K, et al: Interleukin-12 dependent interferon-γ production by CD8α+ lymphoid dendritic cells. J Exp Med 189:1981-1986, 1999

32.   Grewal IS, Foellmer HG, Grewal KD, et al: Requirement for CD40 ligand in costimulation induction, T cell activation, and experimental allergic encephalomyelitis. Science 273:1864-1867, 1996

33.   Green DR, Scott DW: Activation-induced apoptosis in lymphocytes. Curr.Opin.Immunol. 6:476-487, 1994

34.   Watanabe-Fukunaga R, Brannan CI, Copeland NG, Jenkins NA, Nagata S: Lymphoproliferation disorder in mice explained by defects in Fas antigen that mediates apoptosis. Nature 356:314-317, 1992

35.   Nagata S, Suda T: Fas and Fas ligand: lpr and gld mutations. Immunol Today 16:39-43, 1995

36.   Russell JH, Rush B, Weaver C, Wang R: Mature T cells of autoimmune lpr/lpr mice have a defect in antigen-stimulated suicide. Proc Natl Acad Sci USA 90:4409-4413, 1993

37.   Scott DE, Kisch WJ, Steinberg AD: Studies of T cell deletion and T cell anergy following in vivo administration of SEB to normal and lupus-prone mice. J Immunol 150:664-672, 1993

38.   Singer GG, Abbas AK: The Fas antigen is involved in peripheral but not thymic deletion of T lymphocytes in T cell receptor transgenic mice. Immunity 1:365-371, 1994

39.   Kurts C, Heath WR, Kosaka H, Miller JF, Carbone FR: The peripheral deletion of autoreactive CD8+ T cells induced by cross-presentation of self-antigens involves signaling through CD95 (Fas, Apo-1). J Exp Med 188:415-420, 1998

40.   Zhou T, Edwards III CK, Yang P, Wang Z, Bluethmann, H, Mountz JD: Greatly accelerated lymphadenopathy and autoimmune disease in lpr mice lacking tumor necrosis factor receptor I. J Immunol 156:2661-2665, 1996

41.   Zheng L, Fisher G, Miller RE, Peschon J, Lynch DH, Lenardo MJ: Induction of apoptosis in mature T cells by tumor necrosis factor. Nature 377:348-351, 1995

42.   Sytwu, H.-K., Liblau, R.L. & McDevitt, H.O. The roles of Fas/APO-1 (CD95) and TNF in antigen-induced programmed cell death in T cell receptor transgenic mice. Immunity 5:17-30 ,1996

43.   Speiser DE, Sebzda E, Ohteki T, et al: Tumor necrosis factor receptor p55 mediates deletion of peripheral cytotoxic T lymphocytes *in vivo*. Eur J Immunol 26:3055-3060, 1996

44.   Nguyen LT, McKall-Faienza K, Zakarian A, Speiser DE, Mak TW, Ohashi PS: TNF receptor 1 (TNFR1) and CD95 are not required for T cell deletion after virus infection but contribute to peptide-induced deletion under limited conditions. Eur J Immunol 30:683-688, 2000

45.   Zimmermann C, Rawiel M, Blaser C, Kaufmann M, Pircher H: Homeostatic regulation of CD8$^+$ T cells after antigen challenge in the absence of Fas (CD95). Eur J Immunol. 26:2903-2910, 1996

46.   Ehl S, Hoffmann-Rohrer U, Nagata S, Hengartner H, Zinkernagel R: Different susceptibility of cytotoxic T cells to CD95 (Fas/Apo-1) ligand-mediated cell death after activation in vitro versus in vivo. J Immunol 156:2357-2360, 1996

47.   Reich A, Korner H, Sedgwick JD, Pircher H: Immune down-regulation and peripheral deletion of CD8 T cells does not require TNF receptor-ligand interactions nor CD95. Eur J Immunol 30:678-682, 2000

48.   Boise LH, Minn AJ, Noel PJ, et al: CD28 costimulation can promote T cell survival by enhancing the expression of Bcl-x$_L$. Immunity 3:87-98, 1995

49.   Radvanyi L, Shi Y, Vaziri H, et al: CD28 costimulation inhibits TCR-induced apoptosis during a primary T cell response. J Immunol 156:1788-1798, 1996

50.   Vella AT, Mitchell T, Groth B, et al: CD28 engagement and proinflammatory cytokines contribute to T cell expansion and long-term survival in vivo. J Immunol 158:4714-4720, 1997

51.   Coffer PJ, Jin J, Woodgett JR: Protein kinase B (c-Akt): a multifunctional mediator of phosphatidylinositol 3-kinase activation. Biochem J 335:1-13, 1998

52.   Datta SR, Dudek H, Tao X, et al: Akt phosphorylation of BAD couples survival signals to the cell-intrinsic death machinery. Cell 91:231-241, 1997

53.   Jones, R.G., Parsons, M., Bonnard, M., et al. Protein Kinase B (PKB) regulates T lymphocytes survival, Bcl-X$_L$ levels, and NF-$\varkappa$B activation *in vivo*. J Exp Med 191:1709-1720, 2000

54.   Degermann S, Reilly C, Scott B, Ogata L, von Boehmer H, Lo D: On the various manifestations of spontaneous autoimmune diabetes in rodent models. Eur J Immunol 24:3155-3160, 1994

55.   Veis DJ, Sentman CL, Bach EA, Korsmeyer SJ: Expression of the bcl-2 protein in murine and human thymocytes and in peripheral T lymphocytes. J Immunol 151:2546-2554, 1993

56.   Stambolic V, Suzuki A, de la Pompa JL, et al: Negative regulation of PKB/Akt-dependent cell survival by the tumor suppressor PTEN. Cell 95:29-39, 1998

57. Di Cristofano A, Kotsi P, Peng YF, Cordon-Cardo C, Elkon KB, Pandolfi PP: Impaired Fas response and autoimmunity in Pten$^{+/-}$ mice. Science 285:2122-2125, 1999

58. Strasser A, Harris AW, Huang DCS, Krammer PH, Cory S: Bcl-2 and Fas/APO-1 regulate distinct pathways to lymphocyte apoptosis. EMBO J 14:6136-6147, 1995

59. Reap EA, Felix NJ, Wolthuse, A, Kotzin BL, Cohen PL, Eisenberg RA: Bcl-2 transgenic lpr mice show profound enhancement of lymphadenopathy. J Immunol 155:5455-5462, 1995

CONTROL OF AUTOIMMUNITY BY REGULATORY T CELLS

Ethan M. Shevach, Rebecca S. McHugh, Angela M. Thornton, Ciriaco Piccirillo, Kannan Natarajan, and David H. Margulies

Laboratory of Immunology
National Institute of Allergy and Infectious Diseases
National Institutes of Health
Bethesda, MD 20892

1.1 INTRODUCTION

The development of autoimmune disease involves a breakdown in the mechanisms that control self vs non-self discrimimation. The primary mechanism that leads to self tolerance is thymic deletion of autoreactive T cells, but thymic deletion is not perfect and autoreactive T cells do escape to the periphery. Cells that escape thymic deletion are then subject to mechanisms of peripheral tolerance including the induction of anergy[1] as well as T cell ignorance/indifference of the recognition of autoantigens[2]. However, anergy can be reversed and ignorant T cell populations have the potential to be activated when their target self-antigens are released into the lymphoid system during the course of an infectious insult or when activated by cross-reactive antigens present on infectious agents. Passive mechanisms for the induction of self tolerance may therefore be insufficient to control the activation of autoreactive T cells. Evidence has recently been obtained for an active mechanism of immune suppression in which a distinct subset of T cells suppresses the activation of autoreactive T cells that have escaped the other mechanisms of tolerance induction[3]. Two experimental models have been developed which have allowed the definition of unique populations of regulatory T cells. In one model, autoimmunity is induced by depletion of regulatory T cells from adult animals, while in the second model, the development of regulatory T cells is abolished in neonatal animals.

1.2 INDUCTION OF AUTOIMMUNITY BY DEPLETION OF REGULATORY T CELLS

Penhale and colleagues[4] devised procedures which facilitated the depletion of regulatory T cells from adult animals, while leaving autoantigen-specific effector T cells intact. When adult Wistar rats were thymectomized (Tx) and treated with fractional doses of sublethal irradiation, 60% of the treated animals spontaneously developed autoimmune thyroiditis. The active role of regulatory T cells in this model was subsequently demonstrated by preventing the development of autoimmunity by reconstituting the Tx-irradiated rats with lymphoid cells from normal donors after the final dose of irradiation. These experiments strongly suggested that, normally, autoreactive effector and suppressor T cell populations coexist and autoimmunity rarely develops because the balance is in favor of the activity of the suppressor cell populations. In addition to thyroiditis, Penhale et al [5] demonstrated that the Tx-irradiation procedure would also result in the development of autoimmune diabetes in a strain of rats that was normally not susceptible to this disease. No further characterization of the regulatory T cells was carried out until Mason's group[6, 7], a number of years later, used cell surface markers to define the suppressor and effector T cell subset in the rat based on the differential expression of CD45 isoforms. This approach was subsequently extended into the mouse model by Powrie and colleagues[8, 9] who demonstrated that severe inflammatory bowel disease could be induced in immunodeficient SCID mice that were reconstituted with CD4+ CD45RB[high] T cells, while mice that received CD4+CD45RB[low] T cells or a mixture of the two subsets never developed intestinal lesions. Thus, differential expression of CD45RB isoforms allowed a clear separation of autoreactive suppressors and effectors.

The critical role of regulatory T cells in preventing the organ-specific autoimmunity was also demonstrated by preventing their development by subjecting 3 day old mice to Tx. 3dTx mice developed a wide spectrum of organ-specific autoimmune diseases including gastritis, thyroidits, oophoritis, orchitis, pancreatitis and hepatitis[10, 11]. The effector cells in post-3dTx mediated autoimmunity were CD4+ T cells and signs of inflammation and autoantibody production are apparent 3-6 weeks after thymectomy. As mice which were Tx on day 7 of life developed autoimmune disease at a very low incidence, it was originally postulated that a population of suppressor T cells develops slightly later in ontogeny than the autoimmune effector cells and that 3dTx completely abolishes this lineage of cells. The abililty to prevent the induction of disease by reconstitution of 3dTx mice with normal spleen cells by day 10-14 of life was consistent with this hypothesis. The suppressor cells were also shown to be CD4+ T cells and no role for CD8+ T cells could be demonstrated as either effectors or suppressors in this model.

A number of other protocols were developed which resulted in induction of a spectrum of autoimmune diseases which closely resembled that seen post-d3Tx[12]. When newborn mice were treated with cyclosporine A during the first 7 days of life or when adult mice were treated for two weeks, transplantation of their thymuses to *nu/nu* recipients resulted in the development of disease. Co-transfer of spleen cells prevented the induction of autoimmunity. High dose fractionated total lymphoid irradiation of mice also resulted in the development of organ-specific autoimmunity. Neonatal infection of mice with mouse T lymphotropic virus resulted in depletion of CD4+ T cells from the thymus and the subsequent development of autoimmunity. Surprisingly, transgenic (Tg) mice that expressed any rearranged TCR α-chain, but not β-chain gene, under control of the IgH chain enhancer developed T cell-mediated autoimmune disease again resembling the spectrum of autoimmune disease seen in the d3Tx mouse[13]. It is highly likely that those experimental procedures that induced lymphopenia selectively depleted regulatory T cell populations; it is also possible that some of these manipulations may have resulted in a delayed development of the suppressor cell lineage.

1.3   AUTOIMMUNE GASTRITIS AS A MODEL OF ORGAN-SPECIFIC AUTOIMMUNITY

One of the major problems in the analysis of the pathogenesis of any of the models of spontaneous organ-specific autoimmunity is the identification of the initiating target antigen. We elected to focus our studies on the pathogenesis of gastritis which develops after 3dTx of BALB/c mice. This autoimmune gastritis closely resembles pernicious anemia in man[14]. Both are characterized by a lymphocytic infiltrate in the gastric mucosa, parietal cell and chief cell destruction, and circulating autoantibodies against parietal cells, which are primarily specific for the α- and β- chains of the H/K ATPase, the major proton pump of the parietal cell. However, the target antigen for the effector CD4+ T cells had not been defined. When CD4+ T cells were stimulated with purified preparations of the H/K ATPase significant proliferative responses were observed (Table 1). Surprisingly, T cells reactive to the H/K ATPase could only be demonstrated in lymph nodes in the immediate proximity of the stomach, and no response was seen when CD4+ T cells from peripheral lymph nodes were tested[15]. It was likely that the H/K ATPase-reactive T cells were the actual effector cells in this disease, as they could only be detected in mice that developed gastritis, as indicated by the presence of parietal cell specific autoantibodies, gastric inflammation, and the presence of cells capable of transferring disease into *nu/nu* recipients.

Freshly explanted gastric lymph node cells from d3Tx mice responded preferentially to stimulation with purified preparations of the H/K ATPase α-chain[16], and proliferated only marginally to the β-chain (Table 1). To further characterize the fine specificity of anti-

Table 1. Reactivity of T cells from d3Tx mice to H/K ATPase

<u>Antigen</u>	<u>CPM ³H-TdR Incorporation</u>
Control	2,429
H/K ATPase	62,541
α-chain	35,490
β-chain	10,117

H/K ATPase-specific T cells in this model, we generated two cells lines from the gastric lymph nodes of d3Tx mice[17]. Both lines were CD4+, I-A^d restricted, and recognized distinct peptides from the H/K ATPase α-chain (Table 2). One cell line (TXA23) secreted Th1 and the other (TXA51) secreted Th2 cytokines, but both were were equally potent in inducing gastritis. The Th1 cells induced a predominantly lymphocytic infiltrate, while the Th2 cells recruited a mixed infiltrate composed primarily of polymorphonuclear leukocytes and eosinophils. Although the gastritis observed in d3Tx mice is always accompanied by the production of anti-parietal cell antibodies, neither TXA23 or TXA51 cells functioned as helper cells for antbody production in vivo. Furthermore, they were both capable of inducing severe gastritis following transfer to SCID recipients which lack B lymphocytes.

Table 2. Properties of anti-H/K ATPase reactive T cell clones

<u>TXA23</u>	<u>TXA51</u>
Derived from d3Tx LN	Derived from d3Tx LN
Th1--IFN-γ, TNF-α	Th2--IL-4, IL-5, IL-10
TCR-Vα2, Vβ2	TCR- Vα?,Vβ4
Peptide-aa630-641	Peptide-aa889-901
Lymphocytic gastritis	Polymorphonuclear/eosinophilic gastritis

1.3.1 A TCR Transgenic Model of Spontaneous Autoimmune Gastritis

TXA23 cells recognize a peptide from the cytoplasmic region of the H/K ATPase α-chain close to the enzyme catalytic site. Tg mouse lines expressing the TCR expressed by TXA23 were generated. Tg+ TCR expressing cells were positively selected in either H-2^d or H-2^b mice with marked skewing to CD4+ single positive cells only in H-2^d thymocytes. A majority of the peripheral cells expressed the Tg TCR. The positive selection of this receptor in the thymus is somewhat unexpected in view of the high level expression of the H/K ATPase α-chain in the thymus at least at the mRNA level (Figure 1). If high levels of the peptide epitope recognized by the Tg TCR are also present in the thymus, one might have predicted that these cells would have been subject to negative selection. Almost 100% of Tg+ H-2^d, but not H-2^b, animals demonstrated severe gastric pathology. Gastric infiltrates were seen as early as three weeks of age. The Tg+ T cells appeared to be selectively activated within the gastric lymph node as evidenced by a marked increase in CD69 expression (Figure 2), which was not present on Tg+ cells in spleen or lymph node. An increase in CD69+ Tg+ T cells was observed as early as day 10 of life. We are now in the process of evaluating the capacity of the T cells from the Tg animals to transfer disease to both normal immunocompetent animals and *nu/nu* recipients. In preliminary studies, very low numbers (1 x 10^5) of thymocytes, peripheral LN cells, as

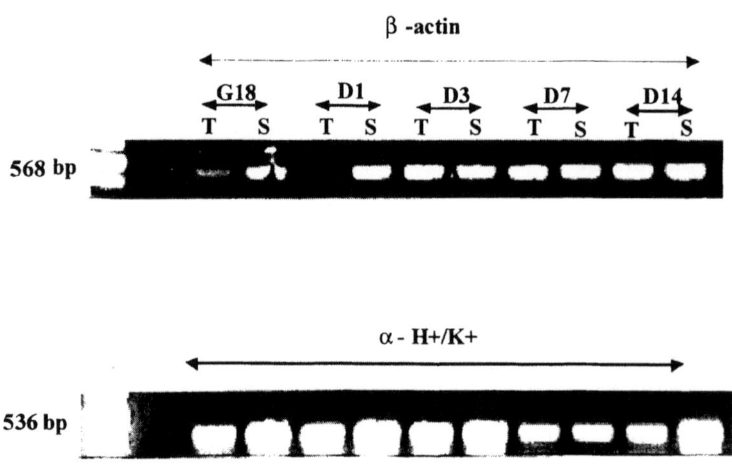

Figure 1. H/K ATPase α-chain gene expression at day 18 of gestation (G18) and early neonatal life.

well as gastric lymph node cells transferred very severe gastritis to *nu/nu* recipients.

In contrast to most murine TCR Tg models of autoimmunity, the rapid spontaneous development of severe autoimmune disease is very unusual. Most models require immunization with the self-antigen[18] for the disease to become manifest or in the case of TCR Tg NOD[19] mice exhibit only modest acceleration of the normal course of the disease process. There are a number of major differences between this Tg model of autoimmunity and the models of Experimental Allergic Encephalomyelitis (EAE), collagen arthritis[20], and diabetes which have been studied by other groups. First, the target autoantigen is known and RT-PCR expression studies clearly demonstrate expression of the antigen in the target tissue as early as day 18 of fetal life (Figure 1). Secondly, as mentioned above the autoreactive T cells are rapidly exported from the thymus early in life and can be detected in the lymph nodes draining the target organ by day 8-10 of life. Lastly, as discussed in Section 1.2, expression of any TCR Tg α-chain appears to retard the development of the suppressor T cell population[13]. Although we have not yet examined the TXA23 transgenic mice for the presence of suppressor T cells, significant damage to the gastric mucosa has already occurred by 3-4 weeks of age; even if the suppressor cells developed at a later time, it would be unlikely that they would be capable of exerting a protective effect. It should be emphasized that a similar delayed maturation of suppressor T cells would occur in other TCR Tg models of autoimmune disease. Why then do they fail to manifest autoimmune disease spontaneously or with a markedly decreased time course, particularly since strong evidence for the presence of regulatory CD4+ T cells exists[21, 22] in these disease models ? One might speculate that early in life in EAE the effector T cells are not activated by exposure to the target antigen in the periphery and cannot enter the target organ unless activated by immunization or by costimulation via infectious agents or their products. Later in life, when cumulative exposure to infections and other environmental insults might lead to T cell activation and entry into the central nervous sytem, the presence of the regulatory T cell populations would inhibit disease induction. In the TCR Tg models of IDDM in the NOD mouse, the nature of the target antigen is still unknown and its expression may be delayed until regulatory T cell populations have emerged from the thymus. The regulatory cells would be capable of partially inhibiting the activation of the TCR Tg T cells and only modest acceleration of the normal course of the disease in NOD

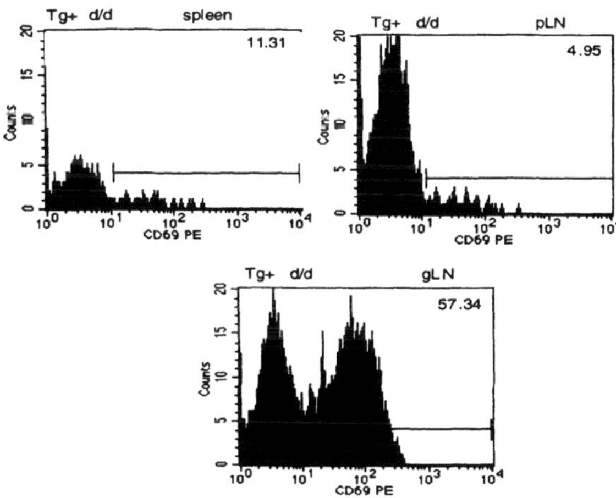

b

Figure 2. Expression of the CD69 antigen on Tg+ T cells in spleen (upper left), peripheral LN (upper right), and gastric LN (lower) in a 6 week old TCR Tg mouse with gastritis. The percentage of CD69+ cells is indicated in the upper right corner of each panel.

mice would be seen. These issues must be addressed in future studies and particular emphasis must be placed as to whether the regulatory CD4+ T cell populations all belong to the same lineage of suppressor T cells (see below).

1.4 CD4+CD25+ T CELLS ARE POTENT IMMUNOREGULATORY CELLS

Although the experiments described above clearly demonstrate that autoimmune disease could be induced by prevention of the development of suppressor T cell populations, a clear implication of all of the studies on regulatory T cells in autoimmunity is that autoimmune effectors and regulatory T cells coexist in the normal host and that depletion of regulatory T cells from the adult animal should result in the development of autoimmune disease. Studies in the rat and mouse demonstrated that the two populations could be separated based on their differential expression of one of the CD45 isoforms[6, 8, 9]. Similarly, Sakaguchi et al[23] demonstrated that effectors and regulatory T cells could be separated based on the relative expression of the Lyt-1(CD5) antigen. Transfer of Lyt-1[low] cells to *nu/nu* mice resulted in the development of autoimmune disease, while transfer of the Lyt-1[high] population did not induce autoimmunity; more importantly, co-transfer of Lyt-1[high] and Lyt-1[low] populations completely prevented the development of disease. Although these studies strongly suggested that the suppressor population could be characterized by high levels of expression of the Lyt-1 antigen, further functional characterization of suppressor T cell function was hampered by the fact that the suppressor cells were present in the large subset (~80%) of the cells that were characterized as Lyt-1[high] based on susceptibility to lysis by antibody and complement. A major deficiency of all of the separation studies was a lack of marker that was exclusively expressed on the suppressor T cell population. Considerable progress toward this goal was provided by the studies of Sakaguchi et al[24, 25] which further defined the suppressor cells as the minor (10%) subset of CD4+ T cells which co-expressed the CD25 (IL-2R α-chain) antigen. Thus, when CD4+ T cells were depleted of this subset and transferred to *nu/nu* recipients, the mice developed a spectrum of autoimmune diseases which closely resembled that observed after 3dTx. Similarly, co-transfer of CD25+ T cells within 10 days of transfer of the CD25- cells completely prevented the development of disease. It is not yet clear that the CD25+ population of suppressor T cells is identical to the suppressor T cell population defined by expression of low levels of the CD45RB antigen although considerable overlap between these two populations undoubtedly exists.

1.4.1 In vivo studies with CD4+ CD25+ T cells

To directly demonstrate that the CD4+CD25+ T cells were solely responsible for suppression of autoimmune disease post-d3Tx[16], we compared the capacity of unseparated normal spleen/LN cells from adult mice or spleen/LN cells depleted of CD25+ T cells to inhibit the development of gastritis post-d3Tx. When d3Tx animals were injected on d10 of life with spleen/LN cells, the development of gastritis was totally abrogated. In contrast, injection of an equal number of spleen/LN from which the CD25+ T cells had been depleted neither diminished nor enhanced the incidence or severity of post-d3Tx induced gastritis. As little information is available of the mechanism by which the CD4+CD25+ T cells inhibit disease in vivo, we attempted to address the question of whether specific cytokines were involved in suppression of disease either post-d3TX or following transfer of CD25- T cells to immunoincompetent recipients. Spleen cells from mice deficient in their capacity to produce IL-4, IL-10, and IFN-γ were as efficient as spleen cells from normal adult BALB/c mice in preventing the induction of disease. Similarly, the protective capacity of normal spleen cells was not reduced when the reconstituted animals were treated with a potent neutralizing antibody to TGF-β strongly suggesting that this suppressor cytokine alone is not responsible for the prevention of autoimmune disease. As NKT cells secrete a number of cytokines which have suppressor functions and NKT cells have been shown to exert protective effects in the NOD mouse model[26], we reconstituted animals with T cells from CD-1 -/- mice which lack NK T cells. Again, the T cells from the CD-1 -/- mice were as efficient as T cells from normal donors in preventing the induction of autoimmune disease. Thus far, we have been unable to further determine whether suppressor cytokines play a role in mediating the disease inhibitory functions of the CD25+ T cell population.

Since CD25 is an activation antigen whose expression can be induced on all CD4+ T cells during the course of an immune response, it remained possible that the CD25+ population was derived from CD4+CD25- T cells that have been activated in vivo in response to normal antigenic stimulation. Alternatively, the CD25+ cells could represent a unique population/lineage of professional immunoregulatory T cells. To address this question, we reconstituted d3Tx mice with a homogeneous population of CD4+ T cells derived from TCR Tg mice on a SCID background which express a single TCR capable of responding to a peptide derived from ovalbumin (OVA)[16]. It should first be noted that these mice have very few CD4+CD25+ T cells as the percentage of CD4 cells which co-express CD25+ is very low (~1%) compared to the percentage (10%) in intact BALB/c mice. This finding has important implications regarding the development of the CD25+ lineage (see below). T cells from the TCR Tg/SCID donor were completely incapable of suppressing disease when used to reconstitute d3Tx mice on day 10 of life. More importantly, activation of the TCR-Tg/SCID T cells in vivo in the reconstitued d3Tx animals by immunization with their target antigen in adjuvant failed to prevent the development of disease even though at least 50% of the Tg T cells in the recipient mice were induced to express CD25. These results are most consistent with the hypothesis that the CD4+CD25+ population in normal mice represents a lineage of cells with unique immunoregulatory properties.

Although CD4+CD25+ T cells were capable of preventing disease when used to reconstitute d3Tx animals on day 10 of life, they could not prevent disease when they were used to reconstitute animals older that 15 days of age. However, it is possible that organ-specific damage had already been initiated by day 15 of life. We were therefore surprised to observe that cloned TXA23 and TXA51 cells could not transfer disease to normal adult BALB/c mice and that co-transfer of CD4+CD25+ T cells with TXA23 cells to *nu/nu* recipients abrogated the capacity of the clones to transfer disease[17]. Conversely, when we transferred the TXA23 clone to TCR Tg/SCID recipients which lacked the CD25+ population, all recipients developed very severe gastritis. It is therefore very likely that the in normal BALB/c mice, the endogenous population of CD4+CD25+ T cells was responsible for the prevention of disease. Thus, the CD25+ cells can inhibit the induction of autoimmune disease and under certain circumstances, prior to the initiation of organ damage, the CD25+ cells can also inhibit the effector functions of fully activated autoreactive T cells.

1.4.2 In Vitro Studies of CD4+CD25+ Suppressor T Cell Function

One major problem in studying the mechanism of suppressor T cell function in vivo is that the assays systems are complex and require weeks to months of assessment of disease activity. It is therefore proven difficult to determine the mechanism of action, antigen specificity, or cellular target of the CD4+CD25+ T cells. To analyze the mechanism of action of these cells, we have established an in vitro model system that mimics the function of these cells in vivo[27]. CD4+CD25+ T cells are purified by magnetic separation or by cell sorting and their capacity to respond in vitro is analyzed. Graded numbers of the CD4+CD25+ T cells are then co-cultured with a fixed number of CD4+CD25- T cells in the presence of an excess of antigen-presenting cells (APC). Some of the results of our initial studies are summarized in Table 3. The CD4+CD25+ T cells are

Table 3. Properties of freshly explanted CD4+CD25+ T cells

--do not respond to anti-CD3 or to the combination of anti-CD3+anti-CD28

--proliferate to the combination of anti-CD3+IL-2

--suppress proliferative responses of CD4+CD25- T cells to anti-CD3

--must be activated via the TCR to suppress

--inhibit the induction of IL-2 mRNA in responders with resultant G_1-S arrest

--suppression can be overcome with IL-2 or anti-CD28

--suppression is cell contact-dependent, cytokine-independent

--are heterogeneous for expression of the other memory/activation markers
  CD45RB, CD62L, CD69, and CD38

--induction of CD25 expression on CD25- T cells does not result in induction of
  suppressor activity

both anergic and suppressive. Both culture experiments in transwells and anti-cytokine antibody neutralization studies have confirmed that the suppression is cell contact dependent. The major mechanism by which the CD4+CD25+ T cells inhibit the the proliferative responses of the CD4+CD25- T cells is by blocking the induction of IL-2 production. Similar conclusions have been reached by Sakaguchi et al[28, 29].

A critical issue which remains to be addressed is the nature of the target antigen recognized by the CD4+CD25+ T cells. Our initial studies demonstrated that activation of the CD4+CD25+ T cells was required for suppression to become manifest since CD4+CD25+ cells from normal BALB/c mice would suppress the response of TCR Tg T cells specific for OVA when anti-CD3 was used as the stimulus, but failed to suppress the response when OVA was used as the stimulus. However, when the CD4+CD25+ isolated from TCR Tg mice on a conventional background were stimulated with their target peptide antigen, they suppressed the responses of CD4+CD25- T cells specific for the same peptide. As discussed above, CD4+CD25+ T cells were not detectable in TCR Tg/ SCID. The presence of CD4+CD25+ T cells in TCR Tg mice maintained on a conventional background likely indicates that the true antigenic specificity of the CD4+CD25+ cell is most likely determined by the endogenous TCR α-chain and not by the Tg α-chain. Although this experiment indicates that the CD4+CD25+ populations can be activated through the Tg TCR, it does not really offer insight into the nature of the physiological antigenic ligand recognized by the CD4+CD25+ T cells as this is recognized by the endogenous, but not the Tg , α–chain expressed by these cells.

A second critical issue is the cellular target of the CD4+CD25 + T cell population We initially raised the possibility that the suppressor and effector populations compete at the APC surface for antigen or costimulatory signals. Table 4 summarizes our recent studies which argue against the possibility that the APC is the cellular target of the

> **Table 4.** CD4+CD25+ T cells do not inactivate APC
>
> --CD4+CD25+ T cells do not inhibit the upregulation of the expression of CD80/CD86, CD40, or CD54.
>
> --CD4+CD25+ T cells are effective inhibitors when LPS-blasts are used as APC.
>
> --CD4+CD25+ T cells are effective inhibitors when fixed LPS-blasts are used as APC.
>
> --Suppression cannot be overcome by the addition of a 40-fold excess of LPS-blasts as APC.

CD4+CD25+ T cells[30]. One additional study which also rules out the APC as the target makes use of our ability to suppress the responses of TCR Tg T cells with CD4+CD25+ derived from the same TCR Tg. We mixed (Table 5) CD4+CD25+ cells isolated from mice expressing a Tg TCR specific for one peptide MHC complex, influenza hemagglutinin$_{110\text{-}119}$-I-E^d (HA) with CD4+ T cells from TCR Tg mice which recognized a distinct peptide MHC complex, pigeon cytochrome C$_{88\text{-}104}$-I-E^k (PCC). When stimulated with HA, CD4+CD25+ T cells from HA TCR Tg mice inhibited the response of CD25- HA TCR Tg T cells. More importantly, HA-specific responses could be inhibited to the same extent in the presence of CD4+CD25+ T cells from PCC TCR Tg mice when both PCC and HA were added to the culture. Similarly, the response of PCC specific T cells could be inhibited by both CD4+CD25+ T cells from HA TCR Tg mice as well as by CD4+CD25+ T cells

Table 5. Cross-suppression of antigen-specific T cell proliferation by CD4+CD25+ T cells.

Responders	Suppressors	Antigen	% Suppression
HA TCR Tg	HA CD25+	HA	100
HA TCR Tg	PCC CD25+	HA	0
HA TCR Tg	PCC CD25+	HA+PCC	100
PCC TCR Tg	PCC CD25+	PCC	100
PCC TCR Tg	HA CD25+	PCC	0
PCC TCR Tg	HA CD25+	PCC+HA	70

from PCC Tg mice. Suppression again required the presence of both peptides in the culture which is consistent with the requirement for activation of the suppressor population. Since two separate and MHC disparate APC populations were used in this study, this cross-suppression experiment is not consistent with a model in which the suppressor T cells compete for an antigenic signal or a co-stimulatory signal on the surface of the APC.

In order to futher analyze the specificity of the CD4+CD25+ T cells, we have been attempting to propagate them in tissue culture. As CD4+CD25+ T cells will respond in short term culture to stimulation with anti-CD3 and IL-2, we have further expanded these cells with IL-2 alone for periods of 7-14 days. Such activated, cultured CD4+CD25+ T cells remain completely anergic and will not proliferate when triggered solely through their TCR[28, 30]. More importantly, in contrast to freshly isolated CD4+CD25+ cells, activated CD4+CD25+ are powerful suppressors of the responses of CD4+CD25- T cells to all tested peptide antigens. The activated suppressor cells do not require further stimulation via their TCR to exert their suppressive effects. There was also no apparent MHC restriction on this in vitro suppressor activity. The suppressor capacity of the cultured CD4+ CD25+ T cells was actually enhanced by a factor of 3-5 fold when compared to freshly isolated CD4+CD25+ T cells with suppression routinely observed at suppressor to effector ratios of 1:16. This two stage culture system now allows us to independently assess and separate the requirements for the activation of suppressor cell activity from the requirements for the delivery of suppressor effector activity. The development of non-specific suppressor function did not require seven days of culture, as CD4+CD25+ T cells from normal BALB/c mice cultured with anti-CD3 and IL-2 for as short a period as 48-72 hours were capable of suppressing the response of TCR Tg T cells to peptide antigen. Collectively, these studies demonstrate that CD4+CD25+ T cells require activation via their TCR to become suppressive, but once activated, their suppressor effector function is completely non-specific.

1.5 CONCLUSIONS AND SPECULATIONS

There are a number of critical questions which remain to addressed before we have a complete understanding of the role of regulatory/suppressor T cells in the immune response. An appropriate starting point is where and how these cells are generated. A number of studies have demonstrated that the CD4+CD25+ T cells are generated in the thymus and acquire their suppressive capabilities as they develop into single positive CD4+ thymocytes[29, 31]. Thus, they emerge from the thymus as CD4+CD25+ cells and are already fully competent to exert their suppressive functions without a additional maturation phase in the periphery. However, CD4+CD25+ T cells cannot be identified either in the thymus or peripheral tissues of IL-2 -/- mice strongly suggesting that IL-2 is required for their maturation in the thymus and/or their survival in the periphery. This result is compatible with our own studies suggesting that the CD4+CD25+ population is a unique lineage of cells with requirements for development and survival distinct from conventional CD4+ and/or CD8+ thymocytes.

We have recently presented a hypothesis which we have termed, "The Altered Negative Selection Model," for the development of immunoregulatory T cells in the thymus[32]. We postulate that it is unlikely that the phenomenon of positive/negative selection in the thymus is an all or none process with high affinity autoreactive cells being deleted and low affinity autoreactive cells being allowed to pass through to the periphery. Some autoreactive T cells may encounter their target peptide-MHC complex with a fit of insufficient affinity to permit deletion, yet not weak enough to allow the cells to pass through. The result will be a signal transduction event in the T cells which results in a permanent change in their potential effector functions when they emerge from the thymus. They will have a "permanent disability" which may be manifest as anergy or as an selective inability to produce Th1 cytokines, but an enhanced or predetermined capacity to produce anti-inflammatory cytokines such as IL-4, IL-10, or TGF-β. Other defects may also be present in the regulatory T cells such as the failure to express the CD40L. Thus, multiple subpopulations of regulatory T cells might be created with specialized functions to suppress autoreactive effectors in different anatomic sites. All the inflammatory bowel disease models have demonstrated the requirement for IL-10 as a critical regulator of immune responses to intestinal bacteria at the mucosal interface; thus, it is not surprising that regulatory T cells which produce this cytokine protect the host against the development of autoimmunity at this local site. On the other hand, inflammation in the brain or the pancreas may be susceptible to a completely different control mechanism.

The model described above would account for the development of these regulatory T cells, but it does not address their survival/maintenance in the periphery. An additional critical question is the mechanism by which they home to a specific site during an inflammatory response. Put simply, how do they know when and where to go? A number of studies suggest that in addition to IL-2, regulatory T cells require the presence of their target organ for survival . For example, suppressor T cell populations derived from the periphery of athyroid rats are unable to suppress autoimmune thyroiditis, but are fully capable of suppressing the development of autoimmune diabetes[33]. Thus, the TCR of the suppressor population is apparently specific for an organ-derived antigen. Is the ligand for the suppressor the same as the ligand for the autoimmune effector cell? It need not be as bystander mechanisms for suppression, e.g., suppresor cytokine production, may be operative. Further studies of chemokine receptor expression on the regulatory T cell populations are also warranted. Lastly, it is also appropriate to raise the issue of how the regulatory T cells distinguish between an autoimmune attack and the normal protective immune response generated to a foreign, pathogen-derived, antigen. It is possible that autoreactive effector cells exclusively express low-affinity anti-autoantigen receptors and thereby are more susceptible to downregulatory effects of suppressor T cells, but this concept is controversial.

Although the production of anti-inflammatory suppressor cytokines is undoubtedly one mechanism by which the regulatory T cells exert their functions, our studies with the CD4+CD25+ T cells in the gastritis model favor a unique cell contact dependent mechanism. As noted above, CD4+CD25+ T cells from IL-4 or IL-10 -/- mice are effective in mediating suppressor T cell functions both in vivo and in vitro. In fact, the TXA51 T cell clone which produces large amounts of both IL-4 and IL-10 functions as an autoreactive effector cell and induces a destructive gastritis. Clearly, suppressor cytokines cannot be the entire story. We strongly considered the possibility that the most effective manner in which a cell contact dependent inhibitory mechanism might operate is that the suppressor cell would target the antigen presenting cell which presents the autoantigenic peptide. Although the CD4+CD25+ T cells may function in this manner in vivo, all of our in vitro studies rule out the APC as the target for the regulatory T cells. On the other hand, we have little insight at present how very few regulatory cells might inactivate a 16-20 fold excess of effector cells. Hopefully, co-transfer studies of regulatory T cells with the Tg anti-H/K ATPase effectors will allow us to begin to probe the cellular basis for immune suppressor cell function in vivo.

There is little doubt from the multiple studies in animal models that suppressor T cells play an important role in protecting the host from the development of organ-specific autoimmune diseases including IDDM, EAE, the post-3dTx syndrome, thyroiditis and inflammatory bowel disease. To date, no conclusive experiments demonstrating the existence of suppressor T cells in well studied models of human organ-specific autoimmunity have been presented. In fact, one of the most successful therapies of autoimmune disease in man for the past 50 years has been the use immunosuppressive drugs. However, these drugs may be a "two-edged sword" and downregulate the functions of suppressor T cells as well as effector T cells[12]. A more appropriate approach for the treatment of autoimmunity in the new millenium should involve the development of procedures to enhance the function or number of regulatory T cells.

2. REFERENCES

1. Schwartz RH: A cell culture model for T lymphocyte anergy. Science 248:1349-1356, 1990
2. Miller JFAP, Heath WR: Self-ignorance in the peripheral T-cell pool. Immunol Rev 133:131-150 1993
3. Moller G: Dominant immunological tolerance. Immunol Rev 149:1-243 1996
4. Penhale WJ, Farmer A, Irvine WJ:. Thyroiditis in T cell-depleted rats: influence of strain, radiation dose, adjuvants and antilymphocyte serum. Clin Exp Immunol 21:362-375 1975
5. Penhale WJ, Stumbles PA, Huxtable CR, Sutherland RJ, Pethick DW: Induction of diabetes in PVG/c strain rats by manipulation of the immune system. Autoimmunity 7:169-179 1990
6. Fowell D, Mason D: Evidence that the T cell repertoire of normal rats contains cells with the potential to cause diabetes. Characterization of the CD4+ T cell subset that inhibits this autoimmune potential. J Exp Med 177:627-636 1993
7. Saoudi A, Seddon B, Heath V, Fowell D, Mason D: The physiological role of regulatory T cells in the prevention of autoimmunity: the function of the thymus in the generation of the regulatory T cell subset. Immunol Rev 149:195-216 1996
8. Powrie F, Leach MW, Mauze S, Caddle LB, Coffman RL: Phenotypically distinct subsets of CD4+ T cells induce or protect from chronic intestinal inflammation in C.B.-17 scid mice. Int Immunol 51:1461-1471 1993

9. Powrie F, Correa-Oliveira R, Mauze S, Coffman RL: Regulatory interactions between CD45RB[high] and CD45RB[low] CD4+ T cells are important for the balance between protective and pathogenic cell-mediated immunity. J Exp Med 179:589 1994

10. Nishzuka Y, Sakakura T: Thymus and reproduction; sex linked dysgenesis of the gonad after neonatal thymectomy in mice. Science 166:753-755 1969

11. Fukuma K, Sakaguchi S, Kuribayashi K, Chen W-L, Morishita R, Sekita K, Uchino H, Matsuda T: Immunologic and clinical studies on murine experimental autoimmune gastritis induced by neonatal thymectomy. Gastroenterology 94:274-283 1988

12. Sakaguchi S, Toda M, Asano M, Itoh M, Morse SS, Sakaguchi N: T cell-mediated maintenance of natural self-tolerance: its breakdown as a possible cause of various autoimmune diseases. J Autoimmun 9:211-220 1996

13. Sakaguchi S, Ermak TH, Toda M, Berg LJ, Ho W, Fazekas de St. Groth B, Peterson PA, Sakaguchi N, Davis MM: Induction of autoimmune disease in mice by germline alteration of the T cell receptor gene expression. J Immunol 152:1471-1484 1994

14. Gleeson PA, Toh B-H, Van Driel I: Organ-specific autoimmunity induced by lymphopenia. Immunol Rev 149:97-126 1996

15. Suri-Payer E, Kehn PJ, Cheever AW, Shevach EM: Pathogenesis of post-thymectomy autoimmune gastritis. Identification of anti-H/K adenosine triphosphatase-reactive T cells. J Immunol 157:1799-1805 1996

16. Suri-Payer E, Amar AZ, Thornton AM, Shevach EM: CD4+CD25+ T cells inhibit both the induction and effector function of autoreactive T cells and represent a unique lineage of immunoregulatory cells. J Immunol 160:1212-1218 1998

17. Suri-Payer E, Amar AZ, McHugh R, Nataragan K, Margulies DH, Shevach EM: Post-thymectomy autoimmune gastritis: fine specificity and pathogenicity of anti-H/K ATPase T cells. Eur J Immunol 29:669-677 1999

18. Lafaille JJ, Nagashima K, Katsuki M, Tonegawa S: High incidence of spontaneous autoimmune encephalomyelitis in immunodeficient anti-myelin basic protein T cell receptor transgenic mice. Cell 78:399-408 1994

19. Katz JD, Wang B, Haskins K, Benoist C, Mathis D: Following a diabetogenic T cell from genesis through pathogenesis. Cell 74:1089-1097 1993

20. Osman GE, Cheunsuk A, Allen AE, Chi E, Liggitt HD, Hood LE, Ladiges WC: Expression of a type II collagen-specific TCR transgene accelerates the onset of arthritis in mice. Int Immunol 10:1613-1633 1998

21. Boitard C, Yasunami R, Dardenne M, Bach JF: T cell-mediated inhibition of the transfer of autoimmune diabetes in NOD mice. J Exp Med 169:1669-1680 1989

22. Olivares-Villagomez D, Wang Y, Lafaille JJ: Regulatory CD4+ T cells expressing endogenous T cell receptor chains protect myelin basic protein-specific transgenic mice from spontaneous autoimmune encephalomyelitis. J Exp Med 188:1883-1894 1998

23. Sakaguchi S, Fukuma K, Kuribayashi K, Matsuda T: Organ-specific autoimmune diseases induced in mice by elimination of T cell subset. I. Evidence for the active participation of T cells in natural self-tolerance; deficit of a T cell subset as a possible cause of autoimmune disease. J Exp Med 161:72-87 1985

24. Sakaguchi S, Sakaguchi N, Asano M, Itoh M, Toda M: Immunologic self-tolerance maintained by activated T cells expressing IL-2 receptor alpha-chains (CD25). Breakdown of a single mechanism of self-tolerance causes various autoimmune diseases. J Immunol 155:1151-1164 1995

25. Asano M, Toda M, Sakaguchi N, Sakaguchi S: Autoimmune disease as a consequence of a developmental abnormality of a T cell subpopulation. J Exp Med 184:387-396 1996

26. Hammond KJL, Poulton LD, Palmisano LJ, Silveira PA, Godfrey DI, Baxter AG, α/β-T cell receptor (TCR)+CD4-CD8- (NKT) thymocytes prevent insulin-dependent diabetes mellitus in nonobese diabetic (NOD)/Lt mice by the influence of interleukin(IL)- 4 and /or IL-10. J Exp Med 187:1047-1056 1998

27. Thornton AM, Shevach EM: CD4+CD25+ immunoregulatory T cells suppress polyclonal T cell activation in vitro by inhibiting interleukin 2 production. J Exp Med 188:287-296 1998

28. Takahashi T, Kuniyasu Y, Toda M, Sakaguchi N, Itoh M, Iwata M, Shimizu J, Sakaguchi S: Immunologic self-tolerance maintained by CD25+CD4+ naturally anergic and suppressive T cells; induction of autoimmune disease by breaking their anergic/suppressive state. Int Immunol 10:1969-1980 1998

29. Itoh M, Takahashi T, Sakaguchi N, Kuniyasu Y, Otsuka F, Sakaguchi S: Thymus and

autoimmunity; production of CD25+CD4+ naturally anergic and suppressive T cells as a key function of the thymus in maintaining immunologic self-tolerance. J Immunol 162:5317 1999

30. Thorton AM, Shevach EM: Suppressor effector functionof CD4+CD25+ immunoregulatory T cells is antigen non-specific. J Immunol 164:183-190 2000

31. Papiernik M, Leite de Moraes M, Pontoux C, Vasseur F, Penit C: Regulatory CD4 T cells: expression of IL-2Rα chain, resistance to clonal deletion and IL-2 dependency. Int Immunol 10:371-378 1998

32. Shevach EM: Regulatory T cells in autoimmunity. Ann Rev Immunol 18:423-449 2000

33. Seddon B, Mason D: Peripheral autoantigen induces regulatory T cells that prevent autoimmunity. J Exp Med 189:877-881 1999

AUTOIMMUNITY, SELF-TOLERANCE AND IMMUNE HOMEOSTASIS: FROM WHOLE ANIMAL PHENOTYPES TO MOLECULAR PATHWAYS

Christopher C. Goodnow*~, Richard Glynne#, Srini Akkaraju#,
Jane Rayner*, David Mack@, James I. Healy#, Shirine Chaudhry*,
Lisa Miosge*, Lauren Wilson*, Peter Papathanasiou & Adele Loy*

*ACRF Genetics Laboratory, Medical Genome Centre,
John Curtin School of Medical Research, Australian National University, Canberra, ACT 2601 Australia
#Dept of Microbiology and Immunology,
Stanford University School of Medicine, Stanford, CA 94305
@Affymetrix, Central Expressway, Santa Clara, CA

1. INTRODUCTION

Current therapy for autoimmune disease is based on broad-spectrum immune suppression, rather than specific correction of defective tolerance mechanisms. On the preventive front, we are not yet able to identify individuals at risk of autoimmune disease or predict clinical course. To develop more specific therapeutic and diagnostic tools, we will need a map of the cellular and molecular pathways and genes that underpin immunological self-tolerance, illuminating the points where the process goes wrong and where it can be corrected.

Tolerance has been studied as a whole animal phenomenon for many decades, but cellular mechanisms have been clearly resolved only in the last twelve years. The key advance was in mouse genetics, enabling Ig- or TCR-transgenic mice to be constructed with increased frequency and decreased heterogeneity of antigen-specific clones (eg anti-hen egg lysozyme, HEL, in our experiments), and enabling new self antigens to be expressed in different forms, amounts, and tissue locations (eg HEL). The cellular processes of self tolerance could be systematically dissected in these animals. Rather than a simple decision of clonal deletion or survival, the conclusion yielded is that individual B and T lymphocytes must pass a remarkable series of self-reactivity checkpoints before they form large clones of effector cells.[1]. Several of these checkpoints involve decisions between cell survival and cell death, whether in the bone marrow[2] or thymus,[3] in the T-zone of the spleen and lymph nodes,[4-7] or within germinal centers.[8] Other checkpoints control the migration pattern of the cells,[4, 9] or control their maturation,[2] proliferation[10-12] or effector cell differentiation (Healy JI and Goodnow CC, in preparation).[3, 14, 15]

Mechanisms of Lymphocyte Activation and Immune Regulation VIII
Edited by Sudhir Gupta, Kluwer Academic/Plenum Publishers, 2001

Charting the molecular pathways and genes that underpin these cellular checkpoints has just begun, and mouse genetics and genomics is again shaping up as a key tool. Here we describe several efforts towards that goal.

2. FAS: AN ESSENTIAL STEP IN A PATHWAY TO ABORT T-CELL DEPENDENT ACTIVATION OF ANERGIC B CELLS.

A single recessive mutation in the mouse, *lpr*, is sufficient to cause subclinical autoantibody production and disrupt actively acquired tolerance in B cells in strains such as C57BL/6 that otherwise have little tendency to autoimmunity.[16, 17] This mouse mutant thus provides a powerful entry point to begin charting molecular pathways of tolerance. Advances in genome mapping techniques early this decade made it possible to identify the *lpr* mutation as disrupting expression of the mouse homologue of a human gene, *FAS*, encoding a lymphoid cell death-inducing receptor.[18] Humans with mutated *FAS* also develop a severe childhood autoantibody disorder, indicating that this pathway serves a conserved function in self-tolerance.[19, 20]

What cellular process of tolerance requires an intact Fas gene? Elegant bone marrow chimera studies showed that the B cell lineage is the primary target of the *lpr* mutation for autoantibody production, and that the mutation caused autoantibody production in a cell autonomous fashion.[21] Transgenic reconstitution experiments reinforce this conclusion.[22, 23] To identify which B cell tolerance checkpoints were disrupted by the *lpr* mutation, we scanned different B cell tolerance checkpoints in transgenic animals bred to the C57BL/6-*lpr* mutation. Maturation arrest and clonal deletion of self reactive B cells in the bone marrow proceed normally in B6-*lpr* animals.[17] Deletion of self-reactive B cells by follicular exclusion in the spleen and deletion of self-reactive cells in germinal centers[8] also occurs normally in short term transfer assays when B cells are homozygous for the *lpr* mutation (unpublished observations). By contrast, when anergic B cells present antigen to CD4 T cells in the T zone of mice in short–term transfer assays, the B cells are normally killed after ~2-3 days by a process that is fully disrupted when the B cells are *lpr*-homozygous.[5, 24]

The ability of *lpr* B cells to escape this one checkpoint appears sufficient to allow gradual clonal expansion of self-reactive B cells and autoantibody production by 6 months of age in Ig/HEL C57BL/6 *lpr* mice.[17] The long time-lag for autoantibody production may reflect the fact that most other checkpoints do not depend on Fas, and that a self-reactive B cell must slip through multiple checkpoints before producing a large number of plasma cells. These other checkpoints may be targets of the autoimmune susceptibility background genes in the MRL-strain. Fas also plays a role in T cell homeostasis by bringing about activation-induced cell death, and this second cellular target of *lpr* appears to be required for autoantibody disease in some analyses[25, 26] but not in others.[22]

3. B7.2, IL-4, AND ANTI-APOPTOTIC MOLECULES REGULATE THE DECISION TO ABORT B CELL ACTIVATION THROUGH FAS.

To define the molecular pathway by which Fas selectively aborts self-reactive B cells, we drew upon the model of HEL/anti-HEL transgenic mice. These animals allow molecular comparisons between homogenous populations of naïve B cells, which have not previously encountered HEL antigen, and anergic B cells which have developed in constant stimulation by self HEL.[27] During in vivo interaction with HEL and HEL-specific CD4 T cells, upregulation of Fas protein on the B cells and elimination of the B cell requires CD40L to be made by the T cell, placing CD40L upstream from Fas in the pathway for this tolerance checkpoint.[28]

34

Fas is upregulated equally on naïve and anergic B cells during in vivo interaction with HEL and T cells, yet the naïve cells are not eliminated but instead stimulated in to proliferation.[28] Selective elimination of self reactive B cells at this checkpoint is decided at some point downstream from Fas, at the level of death signaling/effector mechanisms, by integrating differences in signals generated much earlier in the B cell activation sequence. One of the key differences between the early response to foreign HEL and the early or later response to self HEL is the absence in the latter of any induction of the costimulatory molecule, B7.2/CD86.[10] B7.2 protein increases on the surface of naïve B cells by 10-20-fold within 12 hours after foreign antigen binds to B cell antigen receptors.[29] The lack of B7.2 induction on self-tolerant B cells affects the response of interacting CD4 T cells, so that they make little or no cytokines such as IL-2, IL-4, and other mitogenic factors such as OX40, although they make normal amounts of Fas-ligand mRNA.[30, 31] When B7.2 expression is forced in tolerant cells, via a constitutive B7.2 transgene, this is sufficient to trigger the full spectrum of cytokine production by interacting CD4 cells and block Fas-dependent killing of these cells.[31] In vitro studies with isolated tolerant B cells show that IL4 is sufficient to block Fas-induced death,[32] and in vivo findings of systemic autoantibody production in transgenic mice overexpressing IL-4 reinforce the significance of regulating this pathway.[33]

B7.2 induction appears to be one of several early responses regulating Fas-dependent death of B cells. Other genes induced by foreign antigen in naïve B cells can act to block Fas-induced apoptosis within the B cell, independent of effects on the T cell. These include FAIM, a protein of unknown biochemical function,[34] and two members of the bcl-2 antiapoptotic family, Bcl-XL[35] and A1.[36]

4. DIFFERENT GENE RESPONSES TO ANTIGEN IN NAÏVE AND ANERGIC B CELLS COORDINATE THE DECISION BETWEEN PROLIFERATION AND DEATH.

The failure to induce B7.2 and various anti-apoptotic genes in response to self antigen is part of a broad change in the response characteristics of anergic B cells, which also blocks the T cell-independent mitogenic response to BCR stimulation.[10] Each of these defects is quantitative and potentially reversible, for example if self-antigen ceases to engage the BCR[13, 37] or if a foreign antigen engages the BCR with much greater avidity.[10] Cell cycle entry, B7.2, BclXL and FAIM induction nevertheless take place in the 6-30 hour timeframe, as a result of transcriptional responses immediately downstream of BCR signaling.

To illuminate differences in the transcriptional response immediately downstream from the BCR in naïve and anergic B cells, we used Affymetrix high-density DNA arrays to screen 6,500 genes for changes in their expression in the first 1 or 6 hours after stimulation by foreign antigen and during stimulation by self antigen.[38] Because the differences between naïve, early activated, and tolerant cells are of small magnitude during the initial, defining stages of the response (when all are still in G0 of cell cycle), it was important to compare a statistically valid number of replicate cell samples, and to develop statistical methods for array expression experiments.

By stringent statistical criteria, mRNA from only 28 genes our of 6,500 tested were significantly altered in anergic B cells, when compared to matched populations of naïve B cells. Included amongst these were increased expression of mRNAs encoded negative regulatory molecules: the CD72 cell surface protein which recruits SHP-1; the NAB-2 transcriptional corepressor that blocks early growth response genes 1 and 2 (Egr-1, Egr-2); and two negative regulators of calcium/calmodulin signaling. The role of many of the other genes in establishing B cell anergy has yet to be explored. Using pharmacological inhibitors, BCR-induced expression of a number of the anergy-response genes was shown to be triggered by the calcium/calcineurin/NFAT and MEK/ERK signaling pathways. These signaling pathways are continuously activated by self antigen in anergic B cells.[39] Given the continuous stimulation by antigen and activation of these growth response pathways, it is remarkable how few growth response genes are induced and how closely the gene expression profile of anergic B cells resembles naïve B cells that have not been stimulated by antigen at all.

Further insight into the control of anergic B cell responses came from parallel gene chip analysis of the early activation response to HEL antigen in naïve B cells. During the first hour of the activation response, as the B cell is preparing to move into cell cycle, 59 of the 6500 genes tested make a significant change in mRNA abundance: 37 with increases in mRNA and 22 with decreased mRNA. Among these were genes known to be essential or important for the B cell mitogenic responses: the LSIRF, c-myc, and LKLF transcription factors, and the anti-apoptotic bcl-2 family member, A1. These essential elements of the mitogenic response to antigen are completely blocked in anergic cells, explaining how the T cell-dependent and T-independent responses are blocked in these cells. Strikingly, LSIRF, c-myc, and A1 are all induced by the calcium/calcineurin/NFkB pathway, which is itself essential for B cell mitogenic responses to T-dependent and T-independent antigens and which is selectively uncoupled from BCR signaling in anergic cells.[39]

The detailed molecular picture of B cell anergy that emerges is of repression of key mitogenic response genes at multiple levels: by uncoupling of selected BCR signaling pathways (eg NFkB), and through active repression by gene products (eg Nab2) that are themselves induced by sustained BCR signaling through other pathways (eg ERK). The distinct intracellular milieu that results actively reinforces a tolerant state in the cell, diminishes the self-reactive B cell's potential to make a T-cell independent proliferative immune response, and creates the situation needed for Fas to abort the B cell during T cell dependent responses.

5. GENOME WIDE SCREENS FOR NEW COMPONENTS AND PATHWAYS OF IMMUNE HOMEOSTASIS BY ENU MOUSE MUTAGENESIS.

The *lpr* mutation has provided a key bridge to connect a molecular pathway to a cellular checkpoint and whole animal phenotypes of self-tolerance. There are very few other such mutants available. In invertebrate models such as Drosophila, genome-wide screens for loss-of-function mutations affecting key developmental processes has been the key to illuminating the molecular pathways underpinning tissue morphogenesis. In contrast to gene knockout strategies in the mouse, where only a small number of genes can be tested one at a time, phenotype-based screens in chemically mutagenized Drosophila survey loss of function mutations in thousands of genes to reveal those essential for the process under study. When combined with efficient methods to map and identify mutations, this phenotype-driven approach leads directly to the core regulatory steps in a given process. We have begun a genome-wide screen for loss-of-function mutations in genes essential for immune homeostasis, reasoning that these will play key steps in the cellular checkpoints already identified and in others not yet known.

The feasibility of conducting phenotype-driven mutant screens in mice has emerged as a result of three developments:

(1) the discovery that ethylnitrosourea (ENU) is a supermutagen for mouse spermatogonial stem cells[40];

(2) the development of microsattelite markers to facilitate gross and fine mapping of mutations in the mouse;

(3) increasing information on the physical organization of the genome, gridded BAC libraries of large clones, and mouse EST sequence collections.

(4) sequencing of the human and mouse genomes.

We have established that large-scale phenotype driven screens for recessive mouse mutations causing adult onset disorders can indeed be conducted in the mouse, for relatively low cost, and that this approach yields a rich source of relevant mouse mutants even with relatively unsophisticated screening methods.

We constructed a library of mouse mutants in C57BL/6, a key reference strain for immunology and cancer research, and the strain whose ESTs and genome are being sequenced. Male C57BL/6 mice are treated with ENU, using a regime optimised for that strain, to produce random point mutations in spermatogonial stem cells. When mature sperm are formed from these mutagenized stem cells, the males are mated with normal C57BL/6 female mice. Each of the resulting offspring, G1 animals, inherits a unique constellation of ~100 functional mutations on its paternal chromosomes, together with a wild-type set of chromosomes from the mother.

Recessive loss of function mutations are revealed by establishing 200 separate pedigrees. Each pedigree is founded by a single G1 male, carrying an unknown complement of ~100 functional mutations. The founder male is bred first with normal C57BL/6 females, and then with his (G2) daughters. Each G2 daughter has a 50% chance of carrying in heterozygous form any particular mutation present in the founder. When multiple G2 daughters are mated to the founder, the resulting G3 offspring have a 1 in 8 chance of being homozygous for any particular mutation carried by the founder. By screening 25 G3 offspring, on average 3 homozygotes are expected.

In addition to revealing recessive mutations - the majority class - this strategy improves the specificity of screening by requiring that replicate animals with a given phenotype be observed in a pedigree before proceeding to additional work-up. The pedigree structure allows meiotic recombination to segregate interesting mutations from linked embryonic lethal mutations unless the mutations are closely linked. Each G3 mouse screened is likely to be homozygous for 12 functional mutations, allowing many genes to be screened for a given function at once. We anticipate that in rare instances a mutant phenotype will arise through epistatic interactions between two separate mutations. Strains of this type will be valuable models for the epistasis that occurs in human populations: propagating and mapping a two-locus mutant poses no new challenges over simple Mendelian traits, simply a larger sample size.

Over 20 strains with mutations affecting immune homeostasis have been detected by our first high throughput screens of blood in 200 pedigrees. In addition, this library yielded many recessive mutations causing obesity, wasting, dermatitis, skin/hair abnormalities, bone deformities, ataxia and seizures. Many of the immune defects do not correspond to any known immunological mutants, with phenotypes such as hyperactivated T and or B cells, B cell/stem cell lymphomas, or partial B or T cell deficiencies. Several of the mutations result in absence of T cells or T cell lymphomas that could represent known genes and pathways, although this will be clarified rapidly with chromosomal map location and more precise immunological analyses.

Our present approach is that biology and mapping proceed in parallel for each mutation of interest, with a view to identifying the mutated gene by positional candidate approach . Obtaining a 1-5 cM resolution map location for each mutant is relatively low cost and labour, straightforward, and requires two generations of breeding (ie 5-6 months for immediately penetrant phenotypes). A standard outcrossing strategy and microsattelite marker set can be employed for each mutant to provide an economy of scale, and PCR testing of pooled DNA from 20 affected and 20 unaffected individuals allows the number of PCR reactions to be greatly reduced. Mapping with mouse SNPs will lower these costs further in the near future.

Cell biological analysis of each mutant mouse will define the cell type and developmental stage that is the primary target of the mutation. With this in hand, purified cells will be analyzed by mRNA expression profiling, biochemistry, and proteomics to reveal the molecular pathways affected. An economy of scale exists for parallel biological analysis when several mutant strains are isolated affecting the same cell system (eg our panel of mutant strains with T cell abnormalities). Highly specific biological data, combined with a 1-5 cM map location, will in many cases suggest an obvious candidate gene for resequencing (indeed, most immunological relevant mouse mutations, such as *lpr*, motheaten, and xid have been identified this way). The key resource needed will be detailed gene maps of the mouse genome, which will soon be available through mouse genome sequencing and by synteny with the sequenced human genome.

Because of the success of the current C57BL/6 ENU library, we are moving forward with plans to construct future libraries in inbred strains that are already sensitized to autoimmunity or cancer by carrying a mutation or transgenes. By sensitizing the mutagenized stock, it will be possible to identify modifier gene mutations that prevent autoimmune disease or cancer, as well as modifiers that exacerbate these disorders. Genes and molecular pathways identified in the former category are of particular relevance for drug development. All modifier genes and pathways are significant for developing diagnostic markers to predict and subtype human disease.

6. CONCLUSIONS

Mouse genetics, genome-wide mutagenesis, and mouse genomics, coupled with advances in biochemistry and cell biology, open the way to chart the cellular processes and molecular pathways controlling self-tolerance and autoimmune disease. The molecular definition of these pathways and processes will guide a new generation of immunoregulatory drugs and predictive/diagnostic markers to prevent or treat autoimmune disease.

7. REFERENCES

1. C. C. Goodnow, J. G. Cyster, S. B. Hartley, S. E. Bell, M. P. Cooke, J. I. Healy, S. Akkaraju, J. C. Rathmell, S. L. Pogue, K. P. Shokat, Self-tolerance checkpoints in B lymphocyte development, Adv Immunol 59, 279-368(1995). 2. S. B. Hartley, M. P. Cooke, D. A. Fulcher, A. W. Harris, S. Cory, A. Basten, C. C. Goodnow, Elimination of self-reactive B lymphocytes proceeds in two stages: arrested development and cell death, Cell 72, 325-335(1993).

3. K. M. Murphy, A. B. Heimberger, D. Y. Loh, Induction by antigen of intrathymic apoptosis of CD4+CD8+TCRlo thymocytes in vivo, Science 250, 1720-1723(1990).

4. J. G. Cyster, S. B. Hartley, C. C. Goodnow, Competition for follicular niches excludes self-reactive cells from the recirculating B-cell repertoire [see comments], Nature 371, 389-395(1994).

5. J. C. Rathmell, M. P. Cooke, W. Y. Ho, J. Grein, S. E. Townsend, M. M. Davis, C. C. Goodnow, CD95 (Fas)-dependent elimination of self-reactive B cells upon interaction with CD4+ T cells, Nature 376, 181-184(1995).

6. L. Van Parijs, D. A. Peterson, A. K. Abbas, The Fas/Fas ligand pathway and Bcl-2 regulate T cell responses to model self and foreign antigens, Immunity 8, 265-274(1998).

7. L. Zheng, G. Fisher, R. E. Miller, J. Peschon, D. H. Lynch, M. J. Lenardo, Induction of apoptosis in mature T cells by tumour necrosis factor, Nature 377, 348-351(1995).

8. K. M. Shokat, C. C. Goodnow, Antigen-induced B-cell death and elimination during germinal-centre immune responses, Nature 375, 334-338(1995).

9.   E. R. Kearney, K. A. Pape, D. Y. Loh, M. K. Jenkins, Visualization of peptide-specific T cell immunity and peripheral tolerance induction in vivo, Immunity 1, 327-339(1994).

10.   M. P. Cooke, A. W. Heath, K. M. Shokat, Y. Zeng, F. D. Finkelman, P. S. Linsley, M. Howard, C. C. Goodnow, Immunoglobulin signal transduction guides the specificity of B cell-T cell interactions and is blocked in tolerant self-reactive B cells, J. Exp. Med. 179, 425-438(1994).

11.   R. Schwartz, T cell clonal anergy, Curr. Opin. Immunol. 9, 351-357(1997).

12.   S. Akkaraju, W. Y. Ho, D. Leong, K. Canaan, M. M. Davis, C. C. Goodnow, A range of CD4 T cell tolerance: partial inactivation to organ-specific antigen allows nondestructive thyroiditis or insulitis, Immunity 7, 255-271(1997).

13.   C. C. Goodnow, R. Brink, E. Adams, Breakdown of self-tolerance in anergic B lymphocytes, Nature 352, 532-536(1991).

14.   A. O'Garra, L. Steinman, K. Gijbels, CD4+ T-cell subsets in autoimmunity, Curr Opin Immunol 9, 872-883(1997).

15.   S. Cobbold, H. Waldmann, Infectious tolerance, Curr Opin Immunol 10, 518-524(1998).

16.   P. L. Cohen, R. A. Eisenberg, *Lpr* and gld: single gene models of systemic autoimmunity and lymphoproliferative disease, Ann. Rev. Immunol. 9, 243-269(1991).

17.   J. C. Rathmell, C. C. Goodnow, Effects of the *lpr* mutation on elimination and inactivation of self-reactive B cells, J. Immunol. 153, 2831-2842(1994).

18.   R. Watanabe-Fukunaga, C. I. Branna, N. G. Copeland, N. A. Jenkins, S. Nagata, Lymphoproliferation disorder in mice explained by defects in Fas antigen that mediates apoptosis, Nature 356, 314-317(1992).

19.   G. Fisher, F. Rosenberg, S. Straus, J. Dale, L. Middelton, A. Lin, W. Strober, M. Lenardo, J. Puck, Dominant Interfering Fas Gene Mutations Impair Apoptosis in a Human Autoimmune Lymphoproliferative Syndrome, Cell 81, 935-946(1995).

20.   F. Rieux-Laucat, F. Le Deist, C. Hivroz, A. Roberts, K. Debatin, A. Fischer, J. de Villartay, Mutations in Fas Associated with Human Lymphoproliferative Syndrome and Autoimmunity, Science 268, 1347-1349(1995).

21.   E. S. Sobel, T. Katagiri, K. Katagiri, S. C. Morris, P. L. Cohen, R. A. Eisenberg, An intrinsic B cell defect is required for the production of autoantibodies in the *lpr* model of murine systemic autoimmunity, J. Exp. Med. 173, 1441-1449(1991).

22.   H. Fukuyama, M. Adachi, S. Suematsu, K. Miwa, T. Suda, N. Yoshida, S. Nagata, Transgenic expression of Fas in T cells blocks lymphoproliferation but not autoimmune disease in MRL-*lpr* mice, J Immunol 160, 3805-3811(1998).

23.   H. Komano, Y. Ikegami, M. Yokoyama, R. Suzuki, S. Yonehara, Y. Yamasaki, N. Shinohara, Severe impairment of B cell function in *lpr/lpr* mice expressing transgenic Fas selectively on B cells, Int Immunol 11, 1035-1042(1999).

24.   S. E. Townsend, C. C. Goodnow, Abortive proliferation of rare T cells induced by direct or indirect antigen presentation by rare B cells in vivo, J. Exp Med 187, 1611-1621(1998).

25.   E. S. Sobel, P. L. Cohen, R. A. Eisenberg, *lpr* T cells are necessary for autoantibody production in *lpr* mice, J. Immunol. 150, 4160-4167(1993).

26.   J. Wu, T. Zhou, J. Zhang, J. He, W. Gause, J. Mountz, Correction of accelerated autoimmune disease by early replacement of the mutated *lpr* gene with the normal Fas apoptosis gene in the T cells of transgenic MRL-*lpr/lpr* mice, Proc Natl Acad Sci U S A 15, 2344-2348(1994).

27.   D. Y. Mason, M. Jones, C. C. Goodnow, Development and follicular localization of tolerant B lymphocytes in lysozyme/anti-lysozyme IgM/IgD transgenic mice, International Immunol. 4, 163-175(1992).

28. J. C. Rathmell, S. E. Townsend, J. C. Xu, R. A. Flavell, C. C. Goodnow, Expansion or elimination of B cells in vivo: dual roles for CD40- and Fas (CD95)-ligands modulated by the B cell antigen receptor, Cell 87, 319-329(1996).

29. D. J. Lenschow, A. I. Sperling, M. P. Cooke, G. Freeman, L. Rhee, D. C. Decker, G. Gray, L. M. Nadler, C. C. Goodnow, J. A. Bluestone, Differential up-regulation of the B7-1 and B7-2 costimulatory molecules after Ig receptor engagement by antigen, J Immunol 153, 1990-1997(1994).

30. W. Y. Ho, M. P. Cooke, C. C. Goodnow, M. M. Davis, Resting and anergic B cells are defective in CD28-dependent costimulation of naive CD4+ T cells, J. Exp. Med. 179, 1539-1549(1994).

31. J. C. Rathmell, S. Fournier, B. Weintraub, J. Allison, C. Goodnow, Repression of B7.2 on self-reactive B cells is essential to prevent proliferation and allow Fas-mediated deletion by CD4(+) T cells, J Exp Med 188, 651-659(1998).

32. L. Foote, A. Marshak-Rothstein, T. Rothstein, Tolerant B lymphocytes acquire resistance to Fas-mediated apoptosis after treatment with interleukin 4 but not after treatment with specific antigen unless a surface immunoglobulin threshold is exceeded, J Exp Med 187, 847-853(1998).

33. K. Erb, B. Ruger, M. von Brevern, B. Ryffel, A. Schimpl, K. Rivett, Constitutive expression of interleukin (IL)-4 in vivo causes autoimmune-type disorders in mice, J Exp Med 185, 329-339(1997).

34. T. Schneider, G. Fischer, T. Donohoe, T. Colarusso, R. TL, A novel gene coding for a Fas apoptosis inhibitory molecule (FAIM) isolated from inducibly Fas-resistant B lymphocytes, J Exp Med 189, 949-956(1999).

35. T. Schneider, D. Grillot, L. Foote, G. Nunez, T. Rothstein, Bcl-x protects primary B cells against Fas-mediated apoptosis, J Immunol 159, 4834-4839(1997).

36. R. J. Grumont, I. J. Rourke, S. Gerondakis, Rel-dependent induction of A1 transcription is required to protect B cells from antigen receptor ligation-induced apoptosis, Genes Dev 13, 400-411(1999).

37. J. I. Healy, R. E. Dolmetsch, R. S. Lewis, C. C. Goodnow, Quantitative and qualitative control of antigen receptor signalling in tolerant B lymphocytes, Novartis Found Symp 215, 137-144(1998).

38. R. Glynne, S. Akkaraju, J. I. Healy, J. Rayner, C. C. Goodnow, D. H. Mack, How self-tolerance and the immunosuppressive drug FK506 prevent B-cell mitogenesis, Nature 403, 672-676(2000).

39. J. I. Healy, R. E. Dolmetsch, L. A. Timmerman, J. G. Cyster, M. L. Thomas, G. R. Crabtree, R. S. Lewis, C. C. Goodnow, Different nuclear signals are activated by the B cell receptor during positive versus negative signaling, Immunity 6, 419-428(1997).

40. S. Hitotsumachi, D. A. Carpenter, W. L. Russell, Dose-repetition increases the mutagenic effectiveness of N-ethyl-N-nitrosourea in mouse spermatogonia, Proc natl Acad Sci USA 82, 6619-6621(1985).

PERIPHERAL TOLERANCE AND
ORGAN SPECIFIC AUTOIMMUNITY

Harald von Boehmer and Elmar Jaeckel

Harvard Medical School
Dana-Farber Cancer Institute
Department of Cancer Immunology & AIDS
Boston, MA 02115

1. INTRODUCTION

Concepts of specific immunotolerance were first introduced after the clonal selection theory gained acceptance. In early days Burnet and Lederberg formulated the simple concept that there was a fundamental difference between immature and mature, antigen-receptor bearing lymphocytes such that binding of antigen by the former would result in cell death, while binding to the latter would result in gain of effector function. [1-3] At that time it was not known that immunocytes developed in distinct anatomical sites and thus would perhaps not encounter all self antigens as if they were circulating in the organism like mature lymphocytes. Even then this concept would fail to deal with proteins expressed only in adulthood. Nevertheless, this was the first theory of recessive tolerance that gained experimental support and by now is an established fact. A variant of that theme is the recently discovered receptor-deletion whereby immature B but not T cells change their receptors when confronted with antigen, perhaps because the antigen keeps the rearrangement process ongoing whereby already productive genes are deleted and new ones are being generated. [4,5] Thus we have clonal or cellular deletion as well as receptor editing of immature lymphocytes as possible mechanism that affect immature lymphocytes whereby unresponsiveness is achieved without prior activation to effector function i.e. cytokine secretion, cytolytic activity or antibody production.

The concept of infectious tolerance was coined only later when suppressor lymphocytes or by now regulatory lymphocytes, in fact mostly T cells, were held responsible for preventing responses of naïve T cells [6-8]. Over the years this concept received support from various experimental data [9,10] but it is fair to say that a consensus as to how regulating T cells interfere with immune responses has not yet emerged. While some studies suggest that cytokines such as TGFβ, IL10 or IL4 are major players [11-14] others suggest that soluble factors may have only minor roles and it is really the direct interaction between T cells that matters most, even though here there is little evidence on molecular mechanisms. [15] With regard to tolerance induction in mature T cells we need also be aware of the fact that recessive mechanism that are independent of regulatory T cells do exist: high doses of antigen and chronic stimulation of T cells can result in so-called activation induced cell death whereby lymphocytes die an apoptotic death after a brief phase of effector function. [16-20] Also so-called "anergic" cells may arise after prolonged antigenic stimulation but they may not be as "anergic" as originally thought. [17,21-23]

In spite of central and peripheral tolerance mechanisms autoimmunity in various forms is not exactly an infrequent disease and therefore it is tempting to find out by what mechanisms autoimmune reactions are initiated. Here we have two major schools of thought with regard to organ-specific autoimmune diseases: some authors argue that mechanisms of molecular mimicry are important i.e. that antigens expressed in a tissue-specific fashion are sequestered from the immune system by not being present on antigen-presenting cells like dendritic cells. Autoimmunity would start if microorganisms would elicit immune responses to shared or similar epitopes and activated T cells that exhibit a migratory pattern different from naïve T cells would gain access to the organ and start destruction. This in turn would enhance release of further organ-specific antigenic material that would be presented by cells present in inflammatory tissue. [24-26] One must say that even though this is an appealing concept the experimental evidence supporting it is highly controversial. Other ideas are that organ-specific antigens are normally presented by antigen-presenting cells at low levels insufficient to initiate an immune response but that any non-specific damage to the organ caused by mechanical stress or inflammatory reactions caused by microorganisms infecting preferentially certain tissues could trigger more significant responses and thus cause autoimmune disease. [27]

Obviously these mechanisms are not mutually exclusive but it remains to be determined which are the most significant for any given disease. Thus autoimmune diseases may not represent so much a break in self-tolerance rather than boosting immune response to sequestered self-antigens.

2. RECESSIVE TOLERANCE

2.1. Recessive Central Tolerance

There is no doubt that Burnet and Lederberg were correct with their hypothesis stating that immature lymphocytes would die when confronted with antigen. This has been shown in a variety of experimental systems involving T and B lymphocytes. [28-31] While the mechanism is cell death by apoptosis this is almost all we know with regard to mechanism. Especially the well-established systems of Fas-Fas L mediated apoptosis that work so efficiently in vitro [32] seem to play no essential role in this type of tolerance[33]. Nevertheless, it would appear that the organism relies heavily on this recessive form of tolerance: many antigens, believed to be expressed in an organ-specific fashion outside the thymus were shown to be expressed by specific cells in the thymus and for some of them it was actually shown that they induce tolerance in developing T cells and by doing so avoid certain autoimmune reactions. [34-37] Thus this type of irreversible tolerance that requires the presence of antigen in primary lymphocyte organs represents the most effective way to prevent unwanted immune responses because these immature cells die without having been effector cells.

2.2. Recessive Tolerance in Mature T Cells

Chronic antigenic stimulation with relatively high doses of antigen results in cell death of mature T cells after a brief phase of activation and effector function [16,17,19,20]. Both CD4[+] and CD8[+] T cells are susceptible to this activation induced cell death. It is important to know that while activation of CD8[+] T cells to effector function in vivo and in vitro does not necessarily require CD4$^+$ T cells or CD4$^+$ T cell derived cytokines the CD8$^+$ T cell response can be significantly enhanced and prolonged by the simultaneous activation of CD4$^+$ T cells[38].   Conversely it is much easier to eliminate CD8$^+$ T cells by activation induced cell death in the absence of CD4$^+$ T cell derived factors, especially IL 2[39,40]. Again the essential mechanisms of this type of cell death are unknown as it proceeds unhampered in mice deficient in receptors of the TNF-receptor family i.e. Fas and TNF-RI. [33,41,42] This is not to say that these receptors do not normally contribute but by and large they are not required.

We found that deletion of mature T cells by activation induced cell death is never complete but is as a rule accompanied by the generation of so-called anergic cells i.e. cells that have the relevant antigen-receptor but fail to proliferate when stimulated by antigen. [16,17,21,23] More recent studies have revealed, however, that these cells are not anergic by all criteria. "Anergic" CD4$^+$ and CD8$^+$ T cells were shown to secrete various cytokines some of which, like IL10, are regarded as immuno-suppressive. [21,22] It remains to be seen whether such cells actually function via the secretion of such cytokines or whether they exhibit immunoregulatory functions in other ways like the recently described CD4$^+$25$^+$ cells that appear to suppress immune responses via direct cell contact by unknown molecular mechanisms. [9,15,43]

3. INFECTIOUS OR "DOMINANT" TOLERANCE

The scientific literature on this subject is rather confusing and has not always been reproducible. Nevertheless there are consistent reports that oral administration of antigens leads more often to tolerance than to immunity. [44] Possible mechanisms involve the efficient induction of regulatory T cells that secrete TGF-β as an immunosuppressive cytokine.[45] Deletion of antigen specific T cells has also been invoked. [46] Our own attempts to tolerize CD4$^+$ T cells with transgenic receptor for peptide 111-119 from influenza hemagglutenin by given either peptide or protein orally have failed consistently: not only was there lack of tolerance but systemic activation of T cells could be observed. [17] Thus there may be subtle unknown changes in the oral application of antigen that determine whether tolerance or immunity prevails. At this point in time, however, there is no "safe" protocol that predictably can be used to induce tolerance to a variety of different antigens given via the oral route. [47]

More promising appears the identification of a subset of T cells that leave the thymus of mice several days after birth and were shown to control autoimmune gastritis that develops spontaneously in some strains of mice that are thymectomized shortly after birth. Some of these cells were identified as CD4$^+$ 25$^+$ and were found to be immunosuppressive in several independent co-transfer experiments. [7,9,48] Such cells may secrete IL10 but this may represent a red herring because in vitro studies are claimed to have shown that these cells suppress via direct cell contact with other T cells by an unknown mechanism.[15,43]

It is probably fair to say that cytokines like IL4 or IL10 can modulate immune esponses, depending on the readout employed, but are not the most effective immunosuppressants that entirely prevent immune responses.

4. ORGAN-SPECIFIC AUTOIMMUNE-DISEASE: DIABETES

There exists a series of autoimmune models for diabetes, among them the most commonly used is the NOD mouse that develops diabetes spontaneously. [49-51] Here disease development is well characterized, beginning with peri-insulitis, progressing to invasive insulitis and ending with β-islet cell destruction resulting in diabetes. While this model has been extremely useful in identifying stages of disease and the relevance of potential autoantigens it does not allow to analyze the beginning and the end-phase of the disease in greater detail because the nature of the antigen eliciting the disease as well as the nature of the antigen that is triggering β-cell destruction is unknown as is the specificity the various T cells that mediate the disease process.

To be able to more directly address these questions certain transgenic models of autoimmune diabetes have been produced where T cells express a transgenic αβTCR for natural (unknown) [52] or transgenic (known) islet specific antigens [25,26,53-56]

In one particular case mice expressed on CD8[+] T cells a transgenic TCR specific for a glycoprotein (gp33) from LCMV and gp33 under control of the rat-insulin promoter [25]. Such mice did not develop disease spontaneously but only after inoculation with the LCMV virus that activated the CD8[+] T cells that then destroyed the β-islet cells by a perforin-dependent lytic mechanism. This model may represent an example of molecular mimicry with unknown relevance since the disease process in NOD mice that develop diabetes spontaneously does not essentially depend on perforin [57]. As a rule transgenic disease models where CD4[+] T cells express a transgenic TCR specific for an islet-specific antigen develop diabetes spontaneously without the need for immunization. [52,58] We have constructed such a model where CD4[+] T cells express a transgenic TCR specific for peptide 111-119 from influenza hemagglutenin (HA) and express at the same time HA under the rat insulin promoter. Such mice develop massive insulitis at two weeks of age and proceed to diabetes in 40 to 80 percent (depending on the mouse colony) from 12 weeks of age onwards. Disease progression is associated with increased expression of γ-INF in pancreatic islets by HA-specific T cells, while expression of TNFα, IL4 and IL10 is somewhat but not drastically decreased when mice proceed from insulitis to diabetes. [59]

Being interested in mechanisms that are responsible for the initiation of the disease we constructed a "minimal" model that assumes that islet-specific antigens can gain access to antigen-presenting cells (dendritic cells) of the immune system, activate CD4[+] T cells which in turn cause an inflammatory reaction that destroys specifically β-cells of the pancreas without having to interact with β-islet cells in an antigen-specific manner.[60] Such an experimental system had been designed previously but because of the use of polyspecific T cells was not entirely conclusive with regard to mechanisms that initiate and complete the disease. [54,61] We bred INS-HA expressing mice onto the RAG[-/-] H-2[b] background and reconstituted them under the cover of NK.1 antibodies with bone marrow cells from RAG[-/-] H-2[d] mice. In this way HA protein fragments possibly released from H-2[b] islet cells could be presented by H-2[d] dendritic cells. Such cells should then in turn be able to activate TCR-HA expressing T cells obtained from H-2[d], TCR-HA transgenic, RAG[-/-] mice, i.e. H-2[d] restricted, HA-specific T cells expressing a single αβTCR. Such activated T cells should not be able to interact with H-2[b], HA-expressing β-islet cells and thus would not be able to mediate destruction by a perforin-dependent mechanism. The result of this experiment was that upon injection of the monospecific T cells the H-2[b] INS-HA transgenic, H-2[d] bone marrow reconstituted mice developed diabetes as quickly as H-2[d] INS-HA mice whereas H-2[b] INS-HA mice not reconstituted with H-2[d] bone marrow developed no diabetes at all [60] These data show that at least certain antigens that are expressed in an organ-specific fashion can be released from such organs in the absence of an inflammatory or autoimmune reaction and presented by dendritic cells able to activate autoimmune T cells. They also show that direct antigen-specific contact between CD4[+] T cells and β-islet cells is not required in order to observe diabetes. The fact that such mice were protected form death when injected with insulin indicates that the destruction of β-islet cells in the pancreas was fairly specific and did not involve gross destruction of exocrine pancreatic tissue.

These experiments then raise the fundamental question in which way the β-islet cell destruction is brought about in this particular model as well as in the NOD model. To this end and in order to avoid misleading results in analyzing death receptor gene expression in total islets when using histochemistry or in situ hybridization we analyzed death receptor gene expression by single cell PCR analysis of single β-islet cells expressing preproinsulin. Concentrating on Fas, TNF-R$_1$ and TNF-R$_2$, the latter known to amplify signaling by TNF-R$_1$, we found that all three were significantly upregulated already at the phase of nondestructive insulitis in both the TCR-HA, INS-HA as well as the NOD model. No further increase was observed during the early phase of β-cell destruction when mice became just diabetic but still contained intact β-islet cells. The latter observation may be complicated by the fact that β-islet cells expressing these receptors were rapidly undergoing cell death [62].

The astonishing fact however is that β-cells did not die when they expressed Fas, TNF-R$_1$ and R$_2$ even though the infiltrates contained numerous T cells expressing Fas ligand and T cells as well as other cells expressed TNFα at high levels. These results therefore point to the possibility that β-cell death is specifically regulated through expression as well as modification of gene products interfering with cell death pathways. It is of considerable interest to identify these as they may offer opportunities for therapeutic intervention.

5. CONCLUDING REMARKS

In spite of several potent tolerance mechanisms organ-specific autoimmunity develops due to the fact that normally sequestered autoantigens can gain access to antigen-presenting cells of the immune system leading to T cell activation and organ-specific autoimmune destruction. In certain autoimmune models no molecular mimicry by infectious organisms is required and β-cell destruction does not require killing by perforin-dependent mechanisms that depend on antigen-specific contact between T cells and organ cells. It appears likely that disease progression from insulitis to diabetes is regulated by a decrease of inhibitors of TNF receptors mediated death pathways.

Reference list:

1. F.M.A. Burnet, A modification of Jerne's theory of antibody production using the concept of clonal selection. *Australian Journal of Science* **20,** 67-69 (1957).
2. F.M.A. Burnet, The clonal selection theory. *London Cambridge University Press* (1959).
3. J. Lederberg, Genes and antibodies: Do antigens bear instructions for antibody specificity or they select cell lines that arise by mutation? *Science* **129,** 1649(1959).
4. D. Gay, T. Saunders, S. Camper, M. Weigert, Receptor editing: an approach by autoreactive B cells to escape tolerance. *J Exp Med* **177,** 999-1008 (1993).
5. S.L. Tiegs, D.M. Russell, D. Nemazee, Receptor editing in self-reactive bone marrow B cells. *J Exp Med* **177,** 1009-1020 (1993).
6. R.K. Gershon, K. Kondo, Infectious immunological tolerance. *Immunology* **21,** 903-914 (1971).
7. E.M. Shevach, A. Thornton, E. Suri-Payer, T lymphocyte-mediated control of autoimmunity. *Novartis Found Symp* **215,** 200-211 (1998).
8. R.K. Gershon, K. Kondo, Cell interactions in the induction of tolerance: the role of thymic lymphocytes. *Immunology* **18,** 723-737 (1970).
9. S. Sakaguchi, K. Fukuma, K. Kuribayashi, T. Masuda, Organ-specific autoimmune diseases induced in mice by elimination of T cell subset. I. Evidence for the active participation of T cells in natural self-tolerance: deficit of a T cell subset as a possible cause of autoimmune disease. *J Exp Med* **161,** 72-87 (1985).

10. D. Fowell, A.J. McKnight, F. Powrie, R. Dyke, D. Mason, Subsets of CD4+ T cells and their roles in the induction and prevention of autoimmunity. *Immunol Rev* **123**, 37-64 (1991).

11. A. Miller, A. al-Sabbagh, L.M. Santos, M.P. Das, H.L. Weiner, Epitopes of myelin basic protein that trigger TGF-beta release after oral tolerization are distinct from encephalitogenic epitopes and mediate epitope-driven bystander suppression. *J Immunol* **151**, 7307-7315 (1993).

12. A. Miller, O. Lider, H.L. Weiner, Antigen-driven bystander suppression after oral administration of antigens. *J Exp Med* **174**, 791-798 (1991).

13. Y. Chen, V.K. Kuchroo, J. Inobe, D.A. Hafler, H.L. Weiner, Regulatory T cell clones induced by oral tolerance: suppression of autoimmune encephalomyelitis. *Science* **265**, 1237-1240 (1994).

14. F. Powrie, J. Carlino, M.W. Leach, S. Mauze, R.L. Coffman, A critical role for transforming growth factor-beta but not interleukin 4 in the suppression of T helper type 1-mediated colitis by CD45RB(low) CD4+ T cells. *J Exp Med* **183**, 2669-2674 (1996).

15. A.M. Thornton, E.M. Shevach, CD4+CD25+ immunoregulatory T cells suppress polyclonal T cell activation in vitro by inhibiting interleukin 2 production. *J Exp Med* **188**, 287-296 (1998).

16. B. Rocha, H. von Boehmer, Peripheral selection of the T cell repertoire. *Science* **251**, 1225-1228 (1991).

17. A. Lanoue, C. Bona, H. von Boehmer, A. Sarukhan, Conditions that induce tolerance in mature CD4+ T cells. *J Exp Med* **185**, 405-414 (1997).

18. B. Rocha, A. Grandien, A.A. Freitas, Anergy and exhaustion are independent mechanisms of peripheral T cell tolerance. *J Exp Med* **181**, 993-1003 (1995).

19. S. Webb, C. Morris, J. Sprent, Extrathymic tolerance of mature T cells: clonal elimination as a consequence of immunity. *Cell* **63**, 1249-1256 (1990).

20. C. Kurts, H. Kosaka, F.R. Carbone, J.F. Miller, W.R. Heath, Class I-restricted cross-presentation of exogenous self-antigens leads to deletion of autoreactive CD8(+) T cells. *J Exp Med* **186**, 239-245 (1997).

21. J. Buer, A. Lanoue, A. Franzke, C. Garcia, H. von Boehmer, A. Sarukhan, Interleukin 10 secretion and impaired effector function of major histocompatibility complex class II-restricted T cells anergized in vivo. *J Exp Med* **187**, 177-183 (1998).

22. C. Tanchot, S. Guillaume, J. Delon, C. Bourgeois, A. Franzke, A. Sarukhan, A. Trautmann, B. Rocha, Modifications of CD8+ T cell function during in vivo memory or tolerance induction. *Immunity* **8**, 581-590 (1998).

23. B. Rocha, C. Tanchot, H. von Boehmer, Clonal anergy blocks in vivo growth of mature T cells and can be reversed in the absence of antigen. *J Exp Med* **177**, 1517-1521 (1993).

24. K.W. Wucherpfennig, J.L. Strominger, Molecular mimicry in T cell-mediated autoimmunity: viral peptides activate human T cell clones specific for myelin basic protein. *Cell* **80**, 695-705 (1995).

25. P.S. Ohashi, S. Oehen, K. Buerki, H. Pircher, C.T. Ohashi, B. Odermatt, B. Malissen, R.M. Zinkernagel, H. Hengartner, Ablation of "tolerance" and induction of diabetes by virus infection in viral antigen transgenic mice. *Cell* **65**, 305-317 (1991).

26. M.B. Oldstone, M. Nerenberg, P. Southern, J. Price, H. Lewicki, Virus infection triggers insulin-dependent diabetes mellitus in a transgenic model: role of anti-self (virus) immune response. *Cell* **65**, 319-331 (1991).

27. S.R. Nahill, R.M. Welsh, High frequency of cross-reactive cytotoxic T lymphocytes elicited during the virus-induced polyclonal cytotoxic T lymphocyte response. *J Exp Med* **177**, 317-327 (1993).

28. J.W. Kappler, N. Roehm, P. Marrack, T cell tolerance by clonal elimination in the thymus. *Cell* **49**, 273-280 (1987).

29. H. von Boehmer, P. Kisielow, Self-nonself discrimination by T cells. *Science* **248**, 1369-1373 (1990).

30. P. Kisielow, H. Bluthmann, U.D. Staerz, M. Steinmetz, H. von Boehmer, Tolerance in T-cell-receptor transgenic mice involves deletion of nonmature CD4+8+ thymocytes. *Nature* **333**, 742-746 (1988).

31. S.B. Hartley, J. Crosbie, R. Brink, A.B. Kantor, A. Basten, C.C. Goodnow, Elimination from peripheral lymphoid tissues of self-reactive B lymphocytes recognizing membrane-bound antigens. *Nature* **353**, 765-769 (1991).

32. M. Lenardo, K.M. Chan, F. Hornung, H. McFarland, R. Siegel, J. Wang, L. Zheng, Mature T lymphocyte apoptosis--immune regulation in a dynamic and unpredictable antigenic environment. *Annu Rev Immunol* **17**, 221-253 (1999).

33. G.G. Singer, A.K. Abbas, The fas antigen is involved in peripheral but not thymic deletion of T lymphocytes in T cell receptor transgenic mice. *Immunity* **1**, 365-371 (1994).

34. L. Klein, M. Klugmann, K.A. Nave, B. Kyewski, Shaping of the autoreactive T-cell repertoire by a splice variant of self protein expressed in thymic epithelial cells *Nat Med* **6**, 56-61 (2000).

35. L. Klein, T. Klein, U. Ruther, B. Kyewski, CD4 T cell tolerance to human C-reactive protein, an inducible serum protein, is mediated by medullary thymic epithelium *J Exp Med* **188**, 5-16 (1998).

36. C. Jolicoeur, D. Hanahan, K.M. Smith, T-cell tolerance toward a transgenic beta-cell antigen and transcription of endogenous pancreatic genes in thymus. *Proc Natl Acad Sci U S A* **91**, 6707-6711 (1994).

37. L. Klein and B. Kyewski, Self-antigen presentation by thymic stromal cells: A subtle division of labor. Curr Opin Immunol (2000).in press

38. J. Kirberg, L. Bruno, H. von Boehmer, CD4+8- help prevents rapid deletion of CD8+ cells after a transient response to antigen. *Eur J Immunol* **23**, 1963-1967 (1993).

39. H. von Boehmer, J. Kirberg, L. Bruno, A. Lanoue and A. Sarukhan, Tolerance induction in mature CD8$^+$ and CD4$^+$ T cells. In (Eds. J. Banchereau, B. Dodet, R. Schwartz and E. Trannoy) pp. 5-20, Elsevier, Paris 1996.

40. C. Kurts, F.R. Carbone, M. Barnden, E. Blanas, J. Allison, W.R. Heath, J.F. Miller, CD4+ T cell help impairs CD8+ T cell deletion induced by cross- presentation of self-antigens and favors autoimmunity. *J Exp Med* **186**, 2057-2062 (1997).

41. L.T. Nguyen, K. McKall-Faienza, A. Zakarian, D.E. Speiser, T.W. Mak, P.S. Ohashi, TNF receptor 1 (TNFR1) and CD95 are not required for T cell deletion after virus infection but contribute to peptide-induced deletion under limited conditions. *Eur J Immunol* **30**, 683-688 (2000).

42. A. Reich, H. Korner, J.D. Sedgwick, H. Pircher, Immune down-regulation and peripheral deletion of CD8 T cells does not require TNF receptor-ligand interactions nor CD95 (Fas, APO-1). *Eur J Immunol* **30**, 678-682 (2000).

43. J.G. Chai, I. Bartok, P. Chandler, S. Vendetti, A. Antoniou, J. Dyson, R. Lechler, Anergic T cells act as suppressor cells in vitro and in vivo. *Eur J Immunol* **29**, 686-692 (1999).

44. H.L. Weiner, Oral tolerance: immune mechanisms and treatment of autoimmune diseases. *Immunol Today* **18**, 335-343 (1997).

45. Y. Chen, V.K. Kuchroo, J. Inobe, D.A. Hafler, H.L. Weiner, Regulatory T cell clones induced by oral tolerance: suppression of autoimmune encephalomyelitis. *Science* **265**, 1237-1240 (1994).

46. Y. Chen, J. Inobe, R. Marks, P. Gonnella, V.K. Kuchroo, H.L. Weiner, Peripheral deletion of antigen-reactive T cells in oral tolerance. *Nature* **376**, 177-180 (1995).

47. E. Blanas, F.R. Carbone, J. Allison, J.F. Miller, W.R. Heath, Induction of autoimmune diabetes by oral administration of autoantigen. *Science* **274,** 1707-1709 (1996).
48. E. Suri-Payer, A.Z. Amar, R. McHugh, K. Natarajan, D.H. Margulies, E.M. Shevach, Post-thymectomy autoimmune gastritis: fine specificity and pathogenicity of anti-H/K ATPase-reactive T cells. *Eur J Immunol* **29,** 669-677 (1999).
49. T.L. Delovitch, B. Singh, The nonobese diabetic mouse as a model of autoimmune diabetes: immune dysregulation gets the NOD. *Immunity* **7,** 727-738 (1997).
50. R. Tisch, H. McDevitt, Insulin-dependent diabetes mellitus. *Cell* **85,** 291-297 (1996).
51. J.F. Bach, D. Mathis, The NOD mouse. *Res Immunol* **148,** 285-286 (1997).
52. J.D. Katz, B. Wang, K. Haskins, C. Benoist, D. Mathis, Following a diabetogenic T cell from genesis through pathogenesis. *Cell* **74,** 1089-1100 (1993).
53. C. Kurts, W.R. Heath, F.R. Carbone, J. Allison, J.F. Miller, H. Kosaka, Constitutive class I-restricted exogenous presentation of self antigens in vivo. *J Exp Med* **184,** 923-930 (1996).
54. D. Lo, C.R. Reilly, B. Scott, R. Liblau, H.O. McDevitt, L.C. Burkly, Antigen-presenting cells in adoptively transferred and spontaneous autoimmune diabetes. *Eur J Immunol* **23,** 1693-1698 (1993).
55. B. Scott, R. Liblau, S. Degermann, L.A. Marconi, L. Ogata, A.J. Caton, H.O. McDevitt, D. Lo, A role for non-MHC genetic polymorphism in susceptibility to spontaneous autoimmunity. *Immunity* **1,** 73-83 (1994).
56. S. Degermann, C. Reilly, B. Scott, L. Ogata, H. von Boehmer, D. Lo, On the various manifestations of spontaneous autoimmune diabetes in rodent models. *Eur J Immunol* **24,** 3155-3160 (1994).
57. D. Kagi, B. Odermatt, P. Seiler, R.M. Zinkernagel, T.W. Mak, H. Hengartner, Reduced incidence and delayed onset of diabetes in perforin-deficient nonobese diabetic mice. *J Exp Med* **186,** 989-997 (1997).
58. S. Degermann, C. Reilly, B. Scott, L. Ogata, H. von Boehmer, D. Lo, On the various manifestations of spontaneous autoimmune diabetes in rodent models. *Eur J Immunol* **24,** 3155-3160 (1994).
59. A. Sarukhan, A. Lanoue, A. Franzke, N. Brousse, J. Buer, H. von Boehmer, Changes in function of antigen-specific lymphocytes correlating with progression towards diabetes in a transgenic model. *EMBO J* **17,** 71-80 (1998).
60. A. Sarukhan, O. Lechner, H. von Boehmer, Autoimmune insulitis and diabetes in the absence of antigen-specific contact between T cells and islet beta-cells. *Eur J Immunol* **29,** 3410-3416 (1999).
61. D.M. LaFace, A.B. Peck, Reciprocal allogeneic bone marrow transplantation between NOD mice and diabetes-nonsusceptible mice associated with transfer and prevention of autoimmune diabetes. *Diabetes* **38,** 894-901 (1989).
62. U. Walter, A. Franzke, A. Sarukhan, C. Zober, H. von Boehmer, J. Buer and O. Lechner, Monitoring gene expression of TNFR family members by beta-cells during development of autoimmune diabetes. Eur J Immunol (2000).in press

AUTOIMMUNE LYMPHOPROLIFERATIVE SYNDROME: TYPES I, II AND BEYOND

Hyung J. Chun,[1,2] Michael J. Lenardo[1]

[1]Laboratory of Immunology, National Institute of Allergy and Infectious Diseases, National Institutes of Health, Bethesda, Maryland 20892
[2]Howard Hughes Medical Institute—National Institutes of Health Research Scholars Program

INTRODUCTION

An essential element in the maintenance of homeostasis of the immune system is the capacity for apoptosis, or programmed cell death. Apoptosis limits the accumulation of lymphocytes, and minimizes reactions against self-antigen that can lead to autoimmunity[1,2]. Lymphocyte apoptosis occurs in at least two major forms: antigen receptor engagement and lymphokine withdrawal. These forms of death are controlled in a negative feedback mechanism termed propriocidal regulation[1,3]. Antigen-mediated death of lymphocytes is regulated by Fas (CD95/APO-1), tumor necrosis factor receptor (TNFR), and related molecules[4-6]. Passive apoptosis via lymphokine withdrawal may result from the cytoplasmic activation of caspases that is closely regulated by the mitochondria and the members of the Bcl-2 family of proteins. Apoptosis of other immune cells such as dendritic cells may also contribute to immune homeostasis[7].

Autoimmune lymphoproliferative syndrome (ALPS) provides novel insights into the mechanisms that regulate lymphocyte homeostasis. ALPS is a rare disease characterized by chronic lymphadenopathy, splenomegaly and autoimmune manifestations that results when genetic abnormalities of lymphocyte apoptosis disrupt propriocidal regulation. ALPS is associated with dominant-interfering mutations in components of the death-signaling pathway. Thus far, mutations in three genes encoding three proteins have been identified in ALPS: Fas receptor, Fas ligand (FasL), and Caspase 10/Mch4/FLICE2 (CASP10)[7-10]. Here we provide an overview of the known genetic basis of ALPS, and describe initial experiments undertaken to characterize those patients with the ALPS phenotype lacking a mutation in any of these genes.

Mechanisms of Lymphocyte Activation and Immune Regulation VIII
Edited by Sudhir Gupta, Kluwer Academic/Plenum Publishers, 2001

CLINICAL FINDINGS OF ALPS

ALPS is a rare disorder of lymphocyte homeostasis and immunological tolerance. The disease is defined by three sets of findings: 1) chronic massive, nonmalignant lymphadenopathy and splenomegaly, 2) elevated levels of T cells with $\alpha\beta$ antigen receptors but lacking both CD4 and CD8 coreceptors, and 3) in vitro defects in lymphocyte apoptosis. ALPS patients may also manifest increases in single positive $\alpha\beta$ T cells, $\gamma\delta$ T cells, and B cells, with characteristic histopathologic changes in the lymph node and spleen that can, in rare instances, progress to lymphoma[11]. ALPS patients show a defect in antigen-induced apoptosis in T cells and a defect in Fas-induced apoptosis in both T and B cells[8, 9, 12]. Nearly all of the patients eventually succumb to autoimmune manifestations, including autoimmune glomerulonephritis, inflammatory polyneuropathy, hemolytic anemia, and idiopathic thrombocytopenic purpura[12]. The autoimmunity typically involves antibody-mediated pathology, with a pronounced skewing of CD4 positive T cells towards a T_H2 phenotype with high levels of circulating interleukin-10[13].

Thus far, inherited mutations in three genes have been found to be associated with ALPS. In ALPS type Ia, the phenotype is associated with inherited mutations in the *APT1* gene on chromosome 10q24.1, encoding the Fas receptor[14]. In ALPS type Ib, lymphadenopathy and systemic lupus erythematosus are associated with inherited mutations in the FasL gene on chromosome 1q23[10]. In mice, the *lpr* and *gld* alleles cause similar diseases due to homozygous recessive mutations in the Fas and FasL genes, respectively[15-17]. The major difference between human patients and the animal model is that mice have multi-organ autoimmune disease with renal involvement, whereas various autoimmune cytopenia predominates in ALPS.

A mutation in the CASP10 gene was recently identified in two kindred with the ALPS phenotype who lack a mutation in either the Fas or FasL genes[7]. Patients with this mutation have been designated as ALPS type II.

ALPS TYPE I

Type Ia

Fas receptor is a 44 kD type-1 membrane glycoprotein in the TNFR superfamily[18, 19]. It is translated from nine exons spanning 26 kb on chromosome 10q24.1 (Figure 1). Of these, the first five exons encode a signal sequence and three extracellular cysteine-rich domains (CRDs) that create a binding site for FasL. Exon 6 encodes the transmembrane domain, and exons 7-9 encode the intracellular domain. An 80 amino acid region of the intracellular domain with homology to a conserved region in TNFR1, DR3, DR4 and DR5 has been identified as the "death domain" essential for apoptosis signaling since mutations in this region abrogated apoptosis induction[20].

In the majority of ALPS type Ia patients, the phenotype is manifested by individuals with heterozygous mutations of the *APT1* gene. Two studies have documented that ALPS with homozygous mutations of *APT1* leads to complete Fas deficiency and severe clinical manifestations[9, 21]. In our cohort of ALPS type Ia patients, mutations have been identified in both the extracellular and the intracellular domains of the Fas receptor (Figure 1). All of these mutants studied to date show severely impaired capacity to mediate apoptosis, and interfere with apoptosis mediated by wildtype Fas receptor in co-transfection experiments[8]. Such findings correlate well with the presence of a heterozygous, rather than a homozygous, mutation in the majority of ALPS type Ia patients.

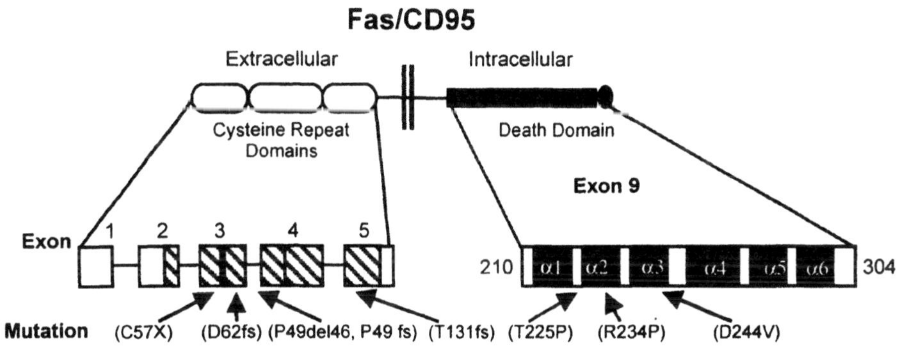

Figure 1. Structure of the Fas receptor. Representative mutations from seven ALPS type Ia patients are shown below the gene diagram. Exon 9 is expanded to show α helical regions of the intracellular death domain.

Among the individuals of ALPS type Ia families, disease penetrance is highly variable. Probands are identified due to significant morbidity, but the same *APT1* mutation could be present in individuals who are clinically normal[8, 12, 22]. Comparison of the intracellular and extracellular mutations have also demonstrated that families with mutations affecting the intracellular domains manifested a higher penetrance of ALPS phenotype when compared to those with mutations affecting the extracellular domains[23].

For the extracellular mutants, our studies have identified interesting mutants that fail to bind to FasL or agonistic antibodies. Such mutants would not be expected to affect trimerization of the wildtype Fas receptors by FasL under the ligand induced trimerization model. However, these mutants dominantly interfered with Fas-mediated apoptosis, to levels similar to those of intracellular mutants. We now have strong evidence that Fas, as well as the other members of the TNFR superfamily, form multimeric complexes on the cell surface in the absence of ligand binding, via a conserved domain we have designated the "Pre-Ligand Association Domain" (PLAD)[24, 25]. The identification of such a domain in the members of the TNFR superfamily offers new directions in understanding receptor function as well as potentially modulating diseases such as ALPS.

Type Ib

Homozygous mutations of FasL in mice result in autoimmune disease and lymphadenopathy, as demonstrated by *gld/gld* mice[16, 17]. In humans, a FasL mutation has been observed in one patient with symptoms of both ALPS and systemic lupus erythematosus (SLE)[10]. Although this patient was noted to have lymphadenopathy that was increased compared to other patients with SLE, the extent of lymphadenopathy in this patient was less than that observed in ALPS type Ia patients. In addition, this patient did not exhibit an increase in $\alpha\beta$ CD4$^-$CD8$^-$ T cells that is a defining characteristic of ALPS type Ia patients. A larger screening of SLE patients would determine the extent of mutant FasL involvement in the progression of SLE and other autoimmune diseases.

Of the patients in the NIH ALPS database, the majority was found to have a mutation in their Fas receptor gene. However, we have been intrigued by those patients with related clinical phenotypes and apoptosis defects, in the absence of either Fas or FasL mutations[12, 26]. The absence of mutations in either Fas or FasL led us to search for other genes involved in the apoptosis signaling pathway that may present as ALPS when defective. In two patients in our database, distinct inherited amino acid substitutions in *CASP10* were identified. These patients presented with autoimmunity and pleotropic apoptosis defects in multiple pathways, but had no molecular abnormalities in Fas, FasL, TNFR1, TNFR2, FADD, or caspase 8 (CASP8). However, one patient had a heterozygous *CASP10* mutation causing a substitution of a leucine by phenylalanine in the p17 subunit of the protease. This mutation was also detected in the patient's mother, but not the father and his two siblings. The second patient had a missense mutation predicted to cause a valine to isoleucine change, also in the p17 subunit. In contrast to the first patient, no wildtype *CASP10* was detected, suggesting that the mutant allele is homozygous. Indeed, examination of the DNA from both parents revealed that each was heterozygous for the mutant allele, indicating that the child had inherited a mutant allele from each parent. Neither of these mutations represented common genetic polymorphisms.

CASP10 belongs to a family of mammalian proteins also known as cysteinyl-aspartate-requiring proteinases that play a critical role in mammalian apoptosis[27-30]. This family of proteins represents the mammalian homologues of the *C. elegans* cell death molecule, CED-3. Caspases are synthesized as inactive zymogens that are activated either via autoprocessing or via the actions of other caspases. One of the caspases that is thought to associate directly with FADD and Fas to form the death-inducing signal complex (DISC) is CASP8[31, 32]. Homozygous targeted disruption of *CASP8* gene leads to impaired heart muscle development and congested accumulation of erythrocytes, resulting in prenatal mortality[33].

The defective CASP10 found in the ALPS type II patient with a heterozygous *CASP10* mutation was found to act in a dominant-negative fashion, as it suppressed apoptosis in cells cotransfected with Fas and mutant CASP10. In addition, the mutant CASP10 also suppressed apoptosis driven via TNFR1, DR3, DR4 (TRAIL-R1), and DR5 (TRAIL-R2) receptors[7]. This protection was found to be specific to death receptor pathways and not other apoptosis pathways, as apoptosis via UV irradiation or staurosporine was not affected. Immunoprecipitation assays performed with antibodies against Fas showed that the mutant CASP10 is corecruited with FADD and CASP8 to the DISCs, suggesting a possible mechanism of apoptosis suppression via dominant inhibition of caspase activation.

Our other patient with ALPS type II inherited two identical defective alleles from heterozygous parents in a classic Mendelian recessive fashion. Transfection assays using both forms of defective CASP10 from the two ALPS type II patients afforded protection against apoptosis, with less severe defect in apoptosis function obtained in transfections with the recessive allele of the second patient.

Abnormalities in Dendritic Cell Homeostasis

Our previous data suggest that T cell-mediated lysis of dendritic cells is mediated by TRAIL[7]. Growth and differentiation of dendritic cells have been extensively characterized, but little is known about their fate after their arrival at the lymphoid organs[34]. In the two ALPS type II patients, direct ligand-dependent cell killing of dendritic cells demonstrated a significantly decreased apoptosis in response to TRAIL compared to healthy controls[7]. These patients also demonstrated large accumulations of CD83-bearing dendritic cells that were confined to the T cell areas of the lymph node. Suppression of dendritic cell death via TRAIL was found to be a phenotype specific to ALPS type II, as ALPS type Ia patients with Fas mutations showed normal apoptosis response to TRAIL and showed no histological alterations. These data suggest that CASP10 plays an important role in the induction of dendritic cell death via TRAIL. They also strongly implicate dendritic cells in the breakdown of immune tolerance, possibly via prolonged stimulation of B and T cells.

ALPS TYPE III

Identification of the two ALPS type II patients with *CASP10* mutations has provided us with a novel genetic defect that further defines the mechanisms of apoptosis and homeostasis of the immune system. However, over thirty patients remain in our database with clinical presentations similar to ALPS, but without detectable Fas mutations. Screening for mutations in FasL and CASP10 are currently in progress. These patients are of great interest, as the genetic determinants of their ALPS phenotype remains a mystery. This group of patients who presents with the ALPS phenotype but do not have a mutation in their *Fas*, *FasL* or *CASP10* genes have been designated as ALPS type III.

Our studies of the ALPS type III patients began at the level of the Fas receptor and the formation of DISC. As previously described by others, DISC formation occurs upon the engagement of Fas by FasL[31, 32, 35, 36, 37]. The activated Fas homotrimer recruits the cytosolic proteins FADD/MORT and the inactive CASP8, to form a complex with the death domain of Fas[32, 37]. Aggregation of CASP8 leads to the autoprocessing and release of active CASP8 into the cytoplasm, leading to a proteolytic cascade and apoptosis[35, 36].

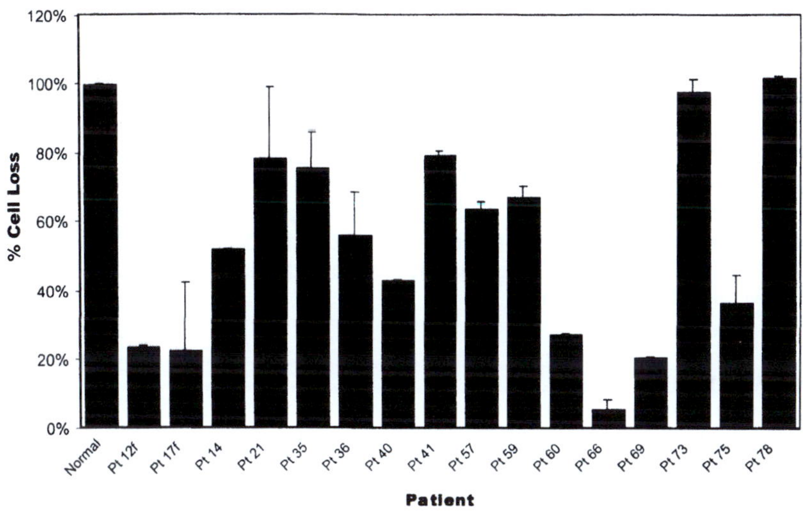

Figure 2. Percent cell loss of EBV transformed B lymphocytes from ALPS type III patients. Cells were treated with 1 µg/ml of APO-1 (Kamiya, WA) for 24 hours.

Our studies also showed that Epstein-Barr Virus transformed B cell lines derived from ALPS type III patients demonstrated a striking variability in their capacity to form a functional DISC. Some of the patients showed a normal DISC formation with cleavage of CASP8 comparable to controls, whereas other ALPS type III patients failed to form a functional DISC, and no active cleaved CASP8 was detected. Such a finding could be explained by a defect in certain patients of the apoptosis signaling pathway that is located upstream of the proteolytic cleavage of CASP8. Other patients who do not show a relative defect in DISC formation would likely have a genetic defect that is located further downstream along the signaling cascade driven by Fas receptor activation.

Our studies of ALPS type II patients suggest that apoptosis pathways in addition to the Fas signaling cascade may be affected to contribute to the phenotype found in ALPS[7]. We subjected the patient lymphocytes to various apoptotic agents, including etoposide, staurosporine, ceramide, and γ–irradiation. Lymphocytes derived from ALPS type III patients showed variable resistance to these agents, as cells from some patients showed protection against apoptosis by specific agents, while others demonstrated apoptosis levels similar to normal controls. These observations suggest means by which the patients can be stratified into disease subtypes that putatively reflect distinct molecular defects. Apoptosis via a number of other death receptors, such as TNFR1 and DR4, are currently being investigated.

The genetic alterations in these patients may provide insights into novel regulatory aspects of antigen-induced apoptosis in human lymphocytes. Recent findings of autoimmune diseases in mouse studies of *PTEN* heterozygous mice and *Bim* deficient mice have provided new avenues to explore the potential mechanisms that may be affected in ALPS type III patients[38, 39]. In our patient population, preliminary analysis of these genes via Western blots showed no significant variation in the expression of these genes. Further analysis of these genes by DNA sequencing will determine any role these genes may play in ALPS.

CONCLUSION

The dilemma in ALPS still remains as to why only a certain subset of patients with the mutation in either their Fas or CASP10 gene suffer the disease, while others remain asymptomatic. Analysis of the ALPS probands and their families suggest that environmental factors or the presence of other immunoregulatory abnormalities are required for full manifestation of the ALPS phenotype. Alternative hypothesis would support the role of protective genes in healthy carriers that counteract the deleterious effects of the mutations. Eventually, the genetic determinants of disease penetrance will be identified, not only allowing for a more precise classification, but also leading to more sophisticated and effective modalities of treatment.

ACKNOWLEDGEMENTS

We thank Stephen Straus, Janet Dale, Jennifer Puck, Roxanne Fischer, Christine Jackson, Lixin Zheng, Richard Siegel, Jin Wang, Francis Chan, Carol Trageser and all the other members of the NIH ALPS team for helpful and interesting collaborative interactions. HJC is supported by the HHMI-NIH Research Scholars Program.

REFERENCES

1. Lenardo MJ: Fas and the art of lymphocyte maintenance. J Exp Med 183: 721-724, 1996
2. Lenardo MJ: Interleukin-2 programs mouse alpha beta T lymphocytes for apoptosis. Nature 353: 858-861, 1991
3. Thompson CB: Apoptosis in the pathogenesis and treatment of disease. Science 267: 1456-1462, 1995
4. Chinnaiyan AM, Dixit VM: Portrait of an executioner: the molecular mechanism of FAS/APO-1- apoptosis. Semin Immunol 9: 69-76, 1997
5. Russell JH, Wang R: Autoimmune gld mutation uncouples suicide and cytokine/proliferation pathways in activated, mature T cells. Eur J Immunol 23: 2379-2382, 1993
6. Zheng L, Fisher G, Miller RE, Peschon J, Lynch DH, Lenardo M: Induction of apoptosis in mature T cells by tumour necrosis factor. Nature 377: 348-351, 1995
7. Wang J, Zheng L, Lobito A, Chan FK, Dale J, Sneller M, Yao X, Puck JM, Straus SE, Lenardo MJ: Inherited human Caspase 10 mutations underlie defective lymphocyte and dendritic cell apoptosis in autoimmune lymphoproliferative syndrome type II. Cell 98: 47-58, 1999
8. Fisher GH, Rosenberg FH, Straus SE, Dale JK, Middleton LA, Lin AY, Strober W, Lenardo MJ, Puck JM: Dominant interfering Fas gene mutations impair apoptosis in a human autoimmune lymphoproliferative syndrome. Cell 81: 935-946, 1995
9. Rieux-Laucat F, Le Deist F, Hivroz C, Roberts IA, Debatin KM, Fischer A, de Villartay JP: Mutations in Fas associated with human lymphoproliferative syndrome and autoimmunity. Science 268: 1347-1349, 1995
10. Wu J, Wilson J, He J, Xiang L, Schur PH, Mountz JD: Fas ligand mutation in a patient with systemic lupus erythematosus and lymphoproliferative disease. J Clin Invest 98: 1107-1113, 1996
11. Lim MS, Straus SE, Dale JK, Fleisher TA, Stetler-Stevenson M, Strober W, Sneller MC, Puck JM, Lenardo MJ, Elenitoba-Johnson KS, Lin AY, Raffeld M, Jaffe ES: Pathological findings in human autoimmune lymphoproliferative syndrome. Am J Pathol 153: 1541-1550, 1998
12. Sneller MC, Wang J, Dale JK, Strober W, Middleton LA, Choi Y, Fleisher TA, Lim MS, Jaffc ES, Puck JM, Lenardo MJ, Straus SE: Clincial, immunologic, and genetic features of an autoimmune lymphoproliferative syndrome associated with abnormal lymphocyte apoptosis. Blood 89: 1341-1348, 1997
13. Fuss IJ, Strober W, Dale JK, Fritz S, Pearlstein GR, Puck JM, Lenardo MJ, Straus SE: Characteristic T helper 2 T cell cytokine abnormalities in autoimmune lymphoproliferative syndrome, a syndrome marked by defective apoptosis and humoral autoimmunity. J Immunol 158: 1912-1918, 1997
14. Straus SE, Sneller M, Lenardo MJ, Puck JM, Strober W: An inherited disorder of lymphocyte apoptosis: the autoimmune lymphoproliferative syndrome. Ann Intern Med 130: 591-601, 1999
15. Watanabe-Fukunaga R, Brannan CI, Copeland NG, Jenkins NA, Nagata S: Lymphoproliferation disorder in mice explained by defects in Fas antigen that mediates apoptosis. Nature 356: 314-317, 1992
16. Lynch DH, Watson ML, Alderson MR, Baum PR, Miller RE, Tough T, Gibson M, Davis-Smith T, Smith CA, Hunter K: The mouse Fas-ligand gene is mutated in gld mice and is part of a TNF family gene cluster. Immunity 1: 131-136, 1994
17. Takahashi T, Tanaka M, Brannan CI, Jenkins NA, Copeland NG, Suda T, Nagata S: Generalized lymphoproliferative disease in mice, caused by a point mutation in the Fas ligand. Cell 76: 969-976, 1994

18. Itoh N, Yonehara S, Ishii A, Yonehara M, Mizushima S, Sameshima M, Hase A, Seto Y, Nagata S: The polypeptide encoded by the cDNA for human cell surface antigen Fas can mediate apoptosis. Cell 66: 233-243, 1991

19. Trauth BC, Klas C, Peters AM, Matzku S, Moller P, Falk W, Debatin KM, Krammer PH: Monoclonal antibody-mediated tumor regression by induction of apoptosis. Science 245: 301-305, 1989

20. Tartaglia LA, Ayres TM, Wong GH, Goeddel DV: A novel domain within the 55 kd TNF receptor signals cell death. Cell 74: 845-853, 1993

21. Kasahara Y, Wada T, Niida Y, Yachie A, Seki H, Ishida Y, Sakai T, Koizumi F, Koizumi S, Miyawaki T, Taniguchi N: Novel Fas (CD95/APO-1) mutations in infants with a lymphoproliferative disorder. Int Immunol 10: 195-202, 1998

22. Infante AJ, Britton HA, DeNapoli T, Middleton LA, Lenardo MJ, Jackson CE, Wang J, Fleisher T, Straus SE, Puck JM: The clinical spectrum in a large kindred with autoimmune lymphoproliferative syndrome caused by a Fas mutation that impairs lymphocyte apoptosis. J Pediatr 133: 629-633, 1998

23. Jackson CE, Britton HA, DeNapoli T, Middleton LA, Lenardo MJ, Jackson CE, Wang J, Fleisher T, Straus SE, Puck JM: Autoimmune Lymphoproliferative Syndrome with Defective Fas: Genotype Influences Penetrance. Am J Hum Genet 64: 1002-1014, 1999

24. Chan F, Chun HJ, Zheng L, Bui K, Siegel RM, Lenardo MJ: A domain in TNF receptors that mediates ligand-independent receptor assembly and signaling. Science, in press

25. Siegel R, Frederiksen JK, Zacharias DA, Chan FK, Johnson M, Lynch D, Tsien RY, Lenardo MJ: Fas preassociation required for apoptosis signaling and dominant inhibition by pathogenic mutations. Science, in press

26. Dianzani U, Bragardo M, DiFranco D, Alliaudi C, Scagni P, Buonfiglio D, Redoglia V, Bonissoni S, Correra A, Dianzani I, Ramenghi U: Deficiency of the Fas apoptosis pathway without Fas gene mutations in pediatric patients with autoimmunity/lymphoproliferation. Blood 89: 2871-2879, 1997

27. Yuan J, Shaham S, Ledoux S, Ellis HM, Horvitz HR: The C. elegans cell death gene ced-3 encodes a protein similar to mammalian interleukin-1 beta-converting enzyme. Cell 75: 641-652, 1993

28. Horvitz HR, Shaham S, Hengartner MO: The genetics of programmed cell death in the nematode Caenorhabditis elegans. Cold Spring Harb Symp Quant Biol 59: 377-385, 1994

29. Alnemri ES, Livingston DJ, Nicholson DW, Salvesen G, Thornberry NA, Wong WW, Yuan J: Human ICE/CED-3 protease nomenclature [letter]. Cell 87: 171, 1996

30. Nicholson DW, Thornberry NA: Caspases: killer proteases. Trends Biochem Sci 22: 299-306, 1997

31. Muzio M, Chinnaiyan AM, Kischkel FC, O'Rourke K, Shevchenko A, Ni J, Scaffidi C, Bretz JD, Zhang M, Gentz R, Mann M, Krammer PH, Peter ME, Dixit VM: FLICE, a novel FADD-homologous ICE/CED-3-like protease, is recruited to the CD95 (Fas/APO-1) death--inducing signaling complex. Cell 85: 817-827, 1996

32. Boldin MP, Varfolomeev EE, Pancer Z, Mett IL, Camonis JH, Wallach D: A Novel Protein That Interacts with the Death Domain of Fas/APO1 Contains a Sequence Motif Related to the Death Domain. J. Biol. Chem. 270: 7795-7798, 1995

33. Varfolomeev EE, Schuchmann M, Luria V, Chiannilkulchai N, Beckmann JS, Mett IL, Rebrikov D, Brodianski VM, Kemper OC, Kollet O, Lapidot T, Soffer D, Sobe T, Avraham KB, Goncharov T, Holtmann H, Lonai P, Wallach D: Targeted disruption of the mouse Caspase 8 gene ablates cell death induction by the TNF receptors, Fas/Apo1, and DR3 and is lethal prenatally. Immunity 9: 267-276, 1998

34. Banchereau J, Steinman RM: Dendritic cells and the control of immunity. Nature 392: 245-252, 1998

35. Kischkel FC, Hellbardt S, Behrmann I, Germer M, Pawlita M, Krammer PH, Peter ME: Cytotoxicity-dependent APO-1 (Fas/CD95)-associated proteins form a death-inducing signaling complex (DISC) with the receptor. Embo J 14: 5579-5588, 1995
36. Medema JP, Scaffidi C, Kischkel FC, Shevchenko A, Mann M, Krammer PH, Peter ME: FLICE is activated by association with the CD95 death-inducing signaling complex (DISC). Embo J 16: 2794-2804, 1997
37. Chinnaiyan AM, O'Rourke K, Tewari M, Dixit VM: FADD, a novel death domain-containing protein, interacts with the death domain of Fas and initiates apoptosis. Cell 81: 505-512, 1995
38. Di Cristofano A, Kotsi P, Peng YF, Cordon-Cardo C, Elkon KB, Pandolfi PP: Impaired Fas response and autoimmunity in Pten+/- mice. Science 285: 2122-2125, 1999
39. Bouillet P, Metcalf D, Huang DC, Tarlinton DM, Kay TW, Kontgen F, Adams JM, Strasser A: Proapoptotic Bcl-2 relative Bim required for certain apoptotic responses, leukocyte homeostasis, and to preclude autoimmunity. Science 286: 1735-1738, 1999

THE ROLE OF MHC CLASS II MOLECULES IN THE PATHOGENESIS AND PREVENTION OF TYPE I DIABETES

Hugh McDevitt, M.D.

Departments of Microbiology and Immunology, and Medicine
Stanford University School of Medicine
Stanford, CA 94305-5402
hughmcd@stanford.edu

1. INTRODUCTION

Although strong associations, and genetic linkage between the Human Leukocyte Antigen (HLA) complex and susceptibility to a wide variety of autoimmune diseases has been documented for the last 30 years, the mechanisms by which genes in the major histocompatibility complex (MHC) mediate susceptibility to autoimmunity remain poorly understood.[1] While the primary functions of the MHC Class I and Class II molecules— antigen presentation of peptides to the receptors on T-cells, and T-cell positive and negative selection in the thymus—are now well documented and at least partially delineated, the precise molecular mechanisms by which particular alleles of MHC Class I or Class II molecules increase or decrease susceptibility to autoimmune diseases have not yet been worked out.

This brief review will deal with two specific aspects of this problem: First, a new approach for analyzing the effect of specific MHC Class II alleles on susceptibility and resistance to Type I diabetes in the non-obese diabetic (NOD) mouse; and second, the application of the information derived from these studies to the development of a safe, non-toxic, epitope-specific immunotherapy to prevent Type I diabetes and/or its recurrence.

2. MECHANISM OF MHC MEDIATED SUSCEPTIBILITY TO TYPE I DIABETES IN THE NOD MOUSE

A detailed analysis of how specific MHC Class II alleles influence susceptibility or resistance to autoimmunity requires a knowledge of the self antigens which are the target of the autoimmune process, and a detailed knowledge of the MHC Class II alleles which are responsible for susceptibility or resistance to the disease under study.

In the non-obese diabetic mouse, there is only one expressed MHC Class II molecule—I-A^{g7}, which is notable for expressing histidine and serine at positions 56 and 57 in the I-A ß chain.[2] Extensive studies in both human populations, and in inbred mice (for references, see reference 1) have shown that the expression of a neutral amino acid at position 57 in the ß chain of either I-A or HLA DQ is associated with susceptibility to Type I diabetes, in concert with numerous other genes not linked to the MHC. In both the NOD mouse, and in humans, MHC-linked genetic susceptibility behaves as a dominant trait with extremely low penetrance. So that most patients express DQ genes with a neutral amino acid (alanine, serine, or valine) in place of aspartic acid at position 57 in both DQ ß chains.

Mechanisms of Lymphocyte Activation and Immune Regulation VIII
Edited by Sudhir Gupta, Kluwer Academic/Plenum Publishers, 2001

2.1. The Critical Target Autoantigens

Numerous ß islet cell proteins have been shown to be the target of either autoantibodies or T-cell proliferative responses in Type I diabetes. In the NOD mouse, the leading candidates are glutamic acid decarboxylase 65 (GAD 65),[3,4] the ß chain residues 9-23 of murine insulin[5] and a peptide (p 277) from murine heat shock protein 60.[6] Numerous studies have shown that these three antigens, GAD 65, the ß chain of insulin, and HSP 60 are all capable of decreasing the incidence and/or delaying the onset of Type I diabetes in the NOD mouse.[7,8,9,10] The administration of each of these three proteins, or peptides derived from them, permits a significant down regulation of the diabeticgenic process. This is one of the principal criteria for their role as the most important targets of the autoimmune process.

Recently, Yoon and his colleagues[11] have presented evidence that prevention of synthesis of GAD 65 (by introduction of a transgene encoding an anti-sense mRNA) in some founder lines can completely suppress diabetes. In these animals, not only is diabetes prevented, but insulitis, and the development of autoantibodies and T-cell proliferative responses to other islet cell proteins is also prevented. If confirmed, this latter result indicates that prevention of the autoimmune response to GAD 65 results in complete abrogation of the autoimmune process in these mice. This suggests that GAD 65 is the most important autoantigen in eliciting the autoimmunity that leads to Type I diabetes. It follows that a detailed analysis of the immune response to GAD 65 in I-A^{g7} NOD mice, with determination of T-cell peptide epitopes and their functional characteristics, is the logical next step in determining how this MHC class II allele mediates susceptibility to Type I diabetes.

2.2. The I-A^{g7} Restricted T-cell Response to GAD 65

In these experiments, 8-12 week old NOD female mice were immunized with GAD 65. GAD 65 was purified from Sf9 insect cell lysates via a 6-histidine tag by elution from a nickel column. The GAD 65 protein was further purified by electroelution from electrophoresis gels. GAD 65 in this form is not enzymatically active and is only partially soluble, requiring treatment with 2M urea to obtain full solubility, followed by dialysis. Mice were immunized with 50 µg GAD 65 emulsified in incomplete Freund's adjuvant, injected into both hind foot pads and over the base of the tail. Ten days later, draining lymph nodes were removed, processed into single-cell suspensions, and cultured *in vitro* with irradiated NOD spleen cells plus GAD 65, 10-20 µg/ml. After three days, the culture fluid was changed and the cells were pulsed overnight with Il-2. The activated T-cells were then fused with a variant of the BW 5147 T-cell tumor which had been mutated to lack expression of its own T-cell receptor and ß chains. The resulting T-cell hybridomas in HAT selective media were plated out by limiting dilution in 96-well plates. Wells showing growth were screened for stimulation of production of Il-2 by exposure to irradiated splenic antigen presenting cells (APC) and GAD 65 10 µg/ml. T-cell hybridomas responding to GAD 65 protein were then screened with 10-12 pools of GAD 65 15-mer peptides spanning the entire GAD sequence (1-585), starting at every fifth residue, thus overlapping by 10 amino acids. T-cell hybridomas responding to a given peptide pool with Il-2 production were then then screened with the individual peptides in the pool of 10 to identify the peptide epitope for which the T-cell hybridoma was specific. The results are shown in Table 1, which presents the GAD 65 peptides recognized by T-cell hybridomas from GAD 65 immunized NOD mice. (This table is the combined result of two separate experiments.)

Table 1. Immunogenic T Cell Epitopes of GAD 65 identified in NOD Mice.

Peptide	Number of T-cell Hybridomas	Percentage (n=74)
206-220	30	41%
221-235	29	39%
286-300	7	9%
401-415	3	4%
561-575	5	7%

To test for the possibility that immunization in incomplete Freund's adjuvant resulted in a skewing of the T-cell repertoire recognizing GAD 65 peptides, an experiment was carried out in which spleen cells from unimmunized 12-week old female NOD mice were used as the fusion partners with BW 5147. (By 12 weeks of age, female NOD mice uniformly have autoantibodies to insulin, GAD 65, and several other islet cell proteins, and show readily detectable T-cell proliferative responses to GAD 65, GAD 67, insulin, carboxypeptidase H, and several other islet cell proteins.) Twelve hundred wells were plated out by limiting dilution in 96-well plates. Growth was observed in 780 wells. All of these were screened for a specific response to GAD 65 protein. Only three wells were stimulated by GAD 65. When tested with individual peptides, one T-cell hybridoma was specific for GAD 65 residues 206-220, and two wells were specific for GAD 65 residues 286-300. Although expensive and time consuming, this experiment tends to validate the results presented in Table 1, since the rare GAD 65 specific T-cells present in spleen were specific for two of the three most common epitopes identified following immunization with GAD 65 protein.[12,13]

A second series of experiments was carried out to study the mechanism by which I-A$_{\beta 57}$ aspartic acid positive I-A molecules confer resistance to Type I diabetes. These experiments were done in mice expressing a site-specific mutated I-A^{g7} ß chain which had been introduced on the NOD background as a transgene. This transgenic mouse (described in reference 13) expresses a single copy of an I-A$_{\beta}^{g7}$ cDNA which has been mutated to express proline and aspartic acid (g7.PD) in positions 56 and 57, in place of the histidine and serine insulin encoded at these positions. Although this transgene was present in a single copy, it led to moderate over-expression of this ß chain. As tested in outcrosses to Balb/c there were approximately two molecules of the transgenic ß chain expressed for every expressed I-A$_{\beta}^{g7}$ wild type ß chain. The results of two separate fusion experiments, carried out by the methods described above, are presented in Table 2. This table shows that in addition to the five GAD 65 epitopes recognized by T-cells in wild type NOD mice, an additional three epitopes were recognized by the T-cell hybridoma repertoire in these transgenic NOD mice. The three new epitopes are GAD 65 456-470, 331-345, and 551-565. It is of interest to note that almost exactly two-thirds of the T-cell hybridomas recognize these new peptide epitopes, closely paralleling the level of expression of the transgenic I-A ß chain.

Table 2. Immunogenic T Cell Epitopes of GAD 65 identified in NOD.PD Transgenic Mice.

Peptide	Number of T-cell Hybridomas	Percentage (n=81)
456-470	42	52%
331-345	5	6%
551-565	8	9%
206-220	3	4%
221-235	2	2%
286-300	11	13%
401-415	3	4%
561-575	7	8%

2.3 MHC Restriction of GAD 65 Specific T-cells in "g7.PD" Mice.

For the hybridomas listed in Table 2, it was necessary to determine their MHC restriction. Accordingly, several hybridomas of each specificity were tested for their ability to respond to the designated GAD 65 peptide presented by antigen presenting cells expressing only I-A^{g7} or I-A$^{g7.PD}$. (The latter antigen presenting cell lines were kindly supplied by Drs. Emil Unanue and Osami Kanagaw, and were produced by transfecting expression constructs for I-A^{g7} or I-A^{g7}.PD into the M.12.C3 cell line, which had been previously mutated to lack expression of all MCH Class II molecules.) The results of these experiments (see reference 12) showed that all T-cell hybridomas with the specificities were listed in Table 1 were restricted to I-A^{g7} APCs, with the exception of hybridomas recognizing GAD 65 221-235. (Some, but not all, of the latter T-cell hybridomas were

Correspondingly, T-cell hybridomas specific for the new peptides listed in Table 2 (the first three peptides in that list) all showed quite sharp restriction to I-A^{g7}.PD, and failed to recognize these peptides when presented by I-A^{g7}. Table 3 presents the amino acid sequence of the 15-mer peptides listed in Tables 1 and 2. The core peptide epitopes, determined by synthesis of peptides truncated from either the amino- or carboxyl-end are indicated for the first three g7 epitopes by the underlining in Table 3.

Table 3. g7 and PD Epitopes and Their Amino Acid Sequences

	Peptide	Sequence
g7	206-220	TYEIAPVFVLLEYVT
epitopes	221-235	LKKMREIIGWPGGSG
	286-300	KKGAAALGIGTDSVI
	401-415	PLQCSALLVREEGLM
	561-575	ISNPAATHGDIDFLI
PD	456-470	WLMWRAKGTTGFEAH
epitopes	331-345	LVSATAGTTVYGAFD
	551-565	GDKVNFFRMVISNPA

2.4. MHC Peptide Binding Affinity of the GAD 65 Peptide Epitopes.

Because earlier experience had shown that I-A^{g7} is relatively unstable in detergents (Fugger, L., Rothbard, J., and McDevitt, H. O., unpublished observations) the relative binding affinity of the g7 and g7.PD epitopes for I-A^{g7} and I-A^{g7}.PD was studied using a cell-binding assay. These results (see reference 12) show that all of the I-A^{g7} epitopes bound to I-A^{g7} with readily measurable IC$_{50}$ (50 percent inhibitory concentration) values ranging from 1-100 μM in this admittedly crude, semiquantitative binding assay. Of the three I-A^{g7}.PD epitopes two (456-470 and 331-345) bound well to I-A^{g7}.PD with IC$_{50}$ values in the 1-10 μM range. A summary of these relative binding values is presented in Table 4. Two points are noteworthy. First, 221-235 binds equally well to I-A^{g7} and I-A^{g7}.PD. Second, differences in binding alone cannot account for the sharp restriction of peptide epitopes to presentation either by I-A^{g7} or I-A^{g7}.PD. For example, peptide 401-415 binds almost equally well to the two I-A alleles, but only elicits T-cells restricted to I-A^{g7}. The same is true to a lesser extent of peptides 456-470 and 331-345 which bind preferentially to g7.PD but also bind somewhat to I-A^{g7}. Despite this, T-cells specific for these latter two peptides are sharply restricted to I-A^{g7}.PD.

Table 4. Relative Binding Capacity of GAD 65 Immunogenic Epitopes to I-A^{g7} and A^{g7}.PD Class II Molecules

Peptide	Immunogenic Epitope of	Binding to I-A^{g7}	Binding to I-A^{g7}.PD
206-220	g7	+++	-
221-235	g7	++++	++++
286-300	g7	++++	+
401-415	g7	++++	+++
561-575	g7	+	+
456-470	PD	+	++++
331-345	PD	+	++++
551-565	PD	-	++

++++, 50% inhibition obtained with 0 to 50 μM inhibitory peptide.
+++, 50% inhibition obtained with 51 to 250 μM inhibitory peptide.
++, 5-% inhibition obtained with 251 to 500 μM inhibitory peptide.
+, 50% inhibition obtained with greater than 501 μM inhibitory peptide
-, no inhibition detected.

2.5. Future Studies.

The studies described above are only the beginning of a detailed analysis of the differential effects of I-A^{g7} and I-A$^{g7.PD}$ on supporting or preventing the development of Type I diabetes in the NOD mouse. More detailed binding studies, including quantitative studies of off-rates for the three major g7 peptide epitopes and the two major g7,PD peptide epitopes are currently underway using a soluble, empty I-A^{g7} molecule produced by Dr. Luc Teyton in drosophila cells. (The DNA constructs encoding these I-A molecules carry positive and negative leucine zippers at the carboxy terminus of the ∂ and ß chains.) Truncation studies to determine the minimal core epitope will be completed for all five peptide. Alanine scans are being done for all five epitopes to determine the MHC and TCR contact residues. Tetramers for the I-A^{g7} molecule binding each of the three principal g7 epitopes have been produced by Dr. Tayton and will be utilized to attempt isolation of T-cells from the islets and pancreatic draining lymph node specific for each of these peptides/MHC complexes. T-cell receptor DNA expression constructs have been produced for T-cell receptors specific for GAD 65 286-300 and 206-220, and are in production for 221-235. These constructs will be used for production of transgenic mice, and for T-cell receptor affinity studies. Subsequently, T-cell receptors specific for the two g7.PD epitopes will also be produced and utilized to produce T-cell receptor transgenic mice.

2.6. Initial Implications

As expected, the T-cell peptide epitopes of GAD 65 recognized by T-cells restricted to I-A^{g7} versus I-A$^{g7.PD}$ are (with one partial exception) different and non-overlapping. The significance of these differences, and the significance of the individual epitopes themselves, will not be apparent until the functional characteristics of T-cells recognizing these peptide/MHC complexes have been analyzed by the methods described above. The functional characteristics of the T-cell populations specific for the g7 epitopes, singly and together, should provide important new insights into how these T-cells contribute to the diabetogenic process. The same is true for functional characteristics of putatively protective T-cells restricted to I-A$^{g7.PD}$ and specific for their respective GAD 65 peptide epitopes.

However, one initial implication can be drawn from these results. The three major peptide epitopes recognized by T-cells restricted to I-A^{g7} in both wild type NOD mice and the I-A$^{g7.PD}$ transgenic mice show readily detectable binding for the I-A^{g7} molecule. Further, utilizing the empty I-A^{g7} molecules referred to above, Dr. Luc Teyton (personal communication) has found that the GAD 65 206-220 peptide has a quite high affinity for I-A^{g7}, with an IC$_{50}$ (concentration for inhibition of binding of a reference peptide) of 80nM. This is a much more direct and reliable semi-quantitative estimate of the affinity of the 206-220 peptide for I-A^{g7}, and indicates that this peptide at least, has a quite high affinity for I-A^{g7}. Examination of Table 3 shows that both 206-220 and 286-300 carry a negatively charged residue (glutamic acid, and aspartic acid) at position P9 and P10 in the peptide, respectively. Several studies[14] have shown that for I-A and HLA DQ molecules lacking aspartic acid at position 57 in the ß chain, peptides with a negatively charged residue at position P9 or P10 in the peptide have a much stronger affinity for these MHC molecules than peptides expressing a neutral or positively charged residue in this position. Thus, it is likely that two of the three principal GAD 65 epitopes eliciting a T-cell response in NOD mice have a relatively high binding affinity for the I-A^{g7} molecule. It would be expected that this higher affinity would result in greater stability, and a longer half life of this peptide/MHC complex on the surface of B-cells and antigen presenting cells.

Correspondingly, these self peptides, as well as other self peptides expressing a negatively charged amino acid at P9 or P10 in the peptide would be expected to be relatively stable in thymic cortical epithelial cells, and would thus be capable of mediating effective negative selection of self reactive T-cells in the thymus. This initial conclusion is somewhat counter to the earlier hypothesis advanced by Unanue and colleagues[15] and Fathman and colleagues[16]. These studies have presented evidence that I-A^{g7} is highly unstable in detergents, is a relatively weak peptide binder (15), and results in readily inducible self reactivity following immunization with peptides closely similar to self peptides (16) (hen egg lysozyme and sperm whale myoglobin peptides). These findings have been adduced to argue that the high degree of self reactivity seen in the NOD mouse strain is due to the relative instability of I-A^{g7}, resulting in inefficient thymic negative selection and the escape to the periphery of a large number of self reactive T-cells.

However, recently published results[17] have shown that the NOD mouse and several other autoimmune strains (NZB, MRL, and BXSB) share an Fc gamma RII gene with a 13 base pair deletion in the 5' promoter region, which results in a striking decrease in expression of Fc gamma RII. This defect, in itself, could be responsible for an increase in autoreactivity, since it has been shown that mice rendered deficient in Fc gamma RII by targeted recombination also show heightened autoreactivity and (in C57BL6) an increased susceptibility to induced autoimmune diseases such as experimental allergic encephalomyelitis (see references in reference 17). Since Fc gamma RII is an inhibitory Fc receptor, and since its deletion results in B-cell and macrophage activation, and increased T-cell self reactivity, it is likely that the heightened degree of autoreactivity seen in the NOD mouse is a result of the defect in Fc gamma RII, and is not due to the postulated instability of I-Ag7. Clearly, this controversy will be subjected to close scrutiny in the coming months.

3. ANTIGEN-SPECIFIC PREVENTION OF TYPE I DIABETES

Numerous approaches to prevention of Type I diabetes, or treatment of acute onset Type I diabetes have been suggested. These include monoclonal antibody therapy with antibodies specific for CD3, CD40 ligand, CD4, adhesion molecules and integens, a variety of methods for introducing Il-4 and Il-10 to down regulate the immune response, and several others (1). Over the coming years, many of these strategies will be subjected to experimental test in human trials, analogous to the ongoing diabetes prevention trial-1 (DPT-1) which utilizes therapy with intact insulin delivered parenterally, or orally, in an attempt to down regulate the diabetic process in individuals who have a permissive HLA type, antibodies to insulin, GAD 65, or IA-2, with or without a demonstrable diminution in insulin reserve.

Each of these approaches will have difficulties and toxicities associated with each therapeutic modality. In the longer term, it might be possible to develop non-toxic, antigen-specific methods of suppressing or down-regulating the immune response to islet cell proteins, thus preventing the complete destruction of ß cells. Numerous preliminary experiments of this sort have been reported over the years (3-10). These have included administration of GAD 65 protein parenterally in pre-diabetic NOD mice (3), administration of the insulin ß chain or peptides from this chain (5), and administration of a peptide derived from heat shock protein 60 (6). Many of these studies have shown that the net effect of delivery of islet cell proteins in a non-inflammatory vehicle (saline, incomplete Freund's ajavent, or alum precipitate) results in a blunting of the T-cell production of T_{H1} cytokines, and an increase in production of T_{H2} cytokines upon stimulation with the specific islet cell protein antigen.

With respect to the peptide epitopes identified above, Tisch has shown that administration of relatively short peptides beginning at residues 217 and 290 of GAD 65, in incomplete Freund's adjuvant to 12-week-old NOD mice results in a significant decrease in the incidence of diabetes at 30 weeks of age.

Adaptation of these results to trials in prediabetic patients is a much more complex matter. First, while 85-90 percent of new Type I diabetic patients are homozygous for lacking aspartic acid on the HLA DQ ß chain, 10-15 percent of these patients have other HLA haplotypes which are aspartic acid positive. Further, the susceptible HLA haplotypes in a U.S. Caucasian population include HLA DR3, 4, and 1. Therefore, the peptide epitopes of insulin, GAD 65, 67, and IA-2 which elicit a T-cell response restricted to HLA DQ B10302 and 0201, and HLA DRB1 0401-0405, 0301, and 0101 need to be defined in detail.[18] Because of the young age of patients with recent onset Type I diabetes, and the difficulty in obtaining repeated large (50 ml.) samples of peripheral blood from the pediatric patient population, alternative approaches to identifying the immunodominant peptide epitopes of these three principal target antigens in human Type I diabetes (18) need to be developed.

One such approach is the utilization of HLA DR and DQ transgenic mice. Such studies are currently underway in a number of laboratories.[19,20,21] These studies have identified immunodominant epitopes of preproinsulin, and GAD 65. These peptide epitopes have been shown to be capable of eliciting a T-cell proliferative response in both recent onset patients with Type I diabetes, and in HLA matched control patients, and to a lesser extent in control patients with other HLA types. Much further analysis of the epitope specificity of the T-cell response to the principal islet cell target antigens needs to be carried out. At the same time vehicles capable of creating a depot of slow release peptides need to be tested for their efficiency, duration and efficacy in animal models and in human trials. Obtaining the necessary preclinical information and preclinical testing will require several years, and considerable investment of research resources and funds. Success in this approach offers the possibility of developing a safe, non-toxic and highly specific means of down-regulating the autoimmune process in patients with pre-Type I diabetes, or in patients receiving islet transplants.

4. REFERENCES

1. R. Tisch and H.O. McDevitt, Insulin-dependent diabetes mellitus, *Cell* **85**, 291-297 (1996).
2.   Acha-Orbea and H.O. McDevitt, The first external domain of the non-obese diabetic mouse class II I-AB chain is unique, *Proc. Natl. Acad. Sci.* **84**(8), 4591-4595 (1987).
3. R.Tisch, X. Yang, S. M. Singer, R. S. Liblau, L. Fugger, and H. O. McDevitt, Immune response to glutamic acid decarboxylase correlates with insulitis onset in non-obese diabetic mice, *Nature* **366**(640), 72-75 (1993).
4. D.L. Kaufman, M. Clare-Saizler, J. Tian, T. Forsthuber, G. S. P. Ting, P. Robinson, M. A. Atkinson, E. E. Sercarz, A. J. Tobin, and P. V. Lehmann, Spontaneous loss of T-cell tolerance to glutamic acid decarboxylase in murine insulin-dependent diabetes, *Nature* **366**(6450), 69-72 (1993).
5. D. Daniel and D. R. Wegmann, Protection of nonobese diabetic mice from diabetes by intranasal or subcutaneous administration of insulin peptide B-(9-23), *Proc. Natl. Acad. Sci.* **93**, 956-960 (1996).
6. I. R. Cohen, Autoimmunity to chaperonins in the pathogenesis of arthritis and diabetes, *Anmu. Rev. Immunol.* **9**:567-589 (1991).
7. R. Tisch, R. S. Liblau, X-D. Yang, P. Liblau, and H. O. McDevitt, Induction of GAD 65-specific regulatory T-cells inhibits ongoing autoimmune diabetes in nonobese diabetic mice, *Diabetes* **47**(6), 1570-1577 (1998).
8. R. Tisch, B. Wang, and D. V. Serreze, Induction of glutamic acid decarboxylase 65-specific Th2 cells and suppression of autoimmune diabetes at late stages of disease is epitope dependent, *J. Immunol.* **163**(3), 1178-1187 (1999).
9.   M. A. Atkinson, N. K. Maclaren, and R. Luchetta, Insulitis and diabetes in NOD mice reduced by prophylactic insulin therapy, *Diabetes* **39**(8), 933-937 (1990).
10. D. Elias, T. Reshef, O. S. Birk, R. van der Zee, M. D. Walker, and I. R. Cohen, Vaccination against autoimmune mouse diabetes with a T-cell epitope of the human 65-kDa heat shock protein, *Proc. Natl. Acad. Sci.* **88**(8), 3088-3091 (1991).
11. J-W. Yoon, C-S. Yoon, H-W. Lim, Q. Q. Huang, Y. Kang, K. H. Pyun, K. Hirasawa, R. S. Sherwin, H-S. Jun, Control of autoimmune diabetes in NOD Mice by GAD expression or suppression in B cells, *Science*, **284**(5417), 1183-1187 (1999).
12. C-C. Chao, H-K. Sytwu, E. L. Chen, J. Toma, and H. O. McDevitt, The role of MHC class II molecules in susceptibility to type I diabetes: identification of peptide epitopes and characterization of the T cell repertoire, *Proc. Natl. Acad. Sci.* **96**(16), 9299-9304 (1999).
13. S. M. Singer, R. Tisch, X-D. Yang, H-K. Sytwu, R. S. Liblau, and H. O. McDevitt, Prevention of diabetes in NOD mice by a mutated I-Ab transgene, *Diabetes* **47**(10), 1570-1577 (1998).
14. G. Rammensee, T. Friede, and S. Stevanoviic, MHC ligands and peptide motifs: first listing, *Immunogenetics* **41**(4), 178-228 (1995).
15. O. Kanagawa, S. M. Martin, B. A. Vaupel, E. Carrasco-Marin, and E. R. Unanue, Autoreactivity of T cells from nonobese diabetic mice: an I-A g7—dependent reaction, *Proc. Natl. Acad. Sci.* **95**(4) , 1721-1724 (1998).
16. W. M. Ridgway, M. Fasso, A. Lanctot, C. Garvey, and C. G. Fathman, Breaking self-tolerance in nonobese diabetic mice, *J. Exp. Med.* **183**(4), 1657-1662 (1996).
17. N. R. Pritchard, A. J. Cutler, S. Uribe, S. J. Chadban, B. J. Morley, and K. G. C. Smith, Autoimmune-prone mice share a promoter haplotype associated with reduced expression and function of the Fc receptor FcyRII, *Curr Biol* **10**:227-230 (2000).
18. O. Rolandsson, E. Hagg, C. Hampe, E. P. Sullivan Jr., M. Nilsson, G. Jansson, G. Hallmans, and A. Lernmark, Glutamate decarboxylase (GAD65) and tyrosine phosphatase-like protein (IA-2) autoantibodies index in a regional population is related to glucose intolerance and body mass index, *Diabetologia* **42**(5), 555-559 (1999).

19. M. Congia, S. Patel, A. P. Cope, S. De Virgiliis, and G. Sonderstrup, T cell epitopes of insulin defined in HLA-DR4 transgenic mice are derived from preproinsulin and proinsulin, *Proc. Natl. Acad. Sci.* **95**(7), 3833-3838 (1998).
20. A. E. Herman, R. Tisch, S. D. Patel, S. L. Parry, J. Olson, J. A. Noble, A. P. Cope, B. Cox, M. Congia, and H. O. McDevitt, Determination of autoantigenic peptides presented by the diabetes-associated HLA-DQ8 class II molecule identifies a motif and responses in individuals with type I diabetes, *J. Immunol.* **163**:6275-6282 (1999).
21. R. S. Abraham and C. S. David, Identification of HLA-class-II-restricted epitopes of autoantigens in transgenic mice, *Curr. Opin. Immunol.* **12**(1), 122-129 (2000).

CONTROL OF AUTOREACTIVE T CELL ACTIVATION BY IMMUNOREGULATORY T CELLS (ART)

Jean-François Bach

INSERM U 25
Hôpital Necker
Paris
France

1. INTRODUCTION

It had been known for several decades that autoimmunity does not necessarily lead to pathological manifestations. In fact, such harmless autoimmunity can be considered as physiologic since it is observed in all healthy individuals, both at the B cell level (natural autoantibodies)[1] and at the T cell level (one can derive MHC class II or organ specific T cell lines from normal peripheral blood lymphocytes).[2] Progression to pathogenic autoimmunity requires autoreactive B and/or T cell activation. This requirement is well illustrated by the classic double transgenic mouse experiment in which coexistence of overexpression of a target antigen (a viral protein) in the β cells of the islets of Langerhans and overexpression of T cells specific for this antigen does not lead to diabetes without activation of the said T cells by infection with the specific virus.[3] The question is thus posed of the origin and the modalities of such activation of autoreactive B or T cells in spontaneously occurring autoimmune diseases. On the other hand, it appears that the progression to clinical manifestations in these diseases is often very slow and preceded by a long phase of preclinical autoimmunity which is apparently more intense than the physiologic autoimmunity mentioned above but yet insufficient to create clinically relevant lesions. This preclinical autoimmune phase is particularly well documented in insulin-dependent diabetes mellitus (IDDM), both in the non-obese diabetic (NOD) mouse and in human disease.[4] The question is to determine what are the mechanisms responsible for such a control of disease progression. Are they related to those preventing undesirable clinical expression of physiological autoimmunity? In this chapter, data will be reviewed that indicate that such mechanisms essentially involve CD4 T cells which begin to be characterized in terms of their phenotype, their autoantigen specificity, their mode of action, their genetic control and their sensitivity to environmental factors.

Mechanisms of Lymphocyte Activation and Immune Regulation VIII
Edited by Sudhir Gupta, Kluwer Academic/Plenum Publishers, 2001

2. MODALITIES OF THE RUPTURE OF IGNORANCE

Ignorance, perhaps better named indifference to allude to the common exposure of autoreactive cells to their specific antigens, requires T cell activation to be broken down. This is achieved in the double transgenic mouse experiment mentioned above by virus-mediated T cell activation.[3] It is performed in experimentally induced autoimmune diseases by sensitization against the target antigen in the presence of adjuvant. The mechanism involved in spontaneously occurring autoimmune diseases is more open to speculation. Five orders of mechanisms can be proposed :

2.1. Target organ inflammation

The selective inflammation of the target organ may lead to the local overexpression of various cytokines or chemokines which attract various effector cells and promote their activation notably through the increased or aberrant expression of the molecules involved in antigen recognition by T cells (e.g. MHC and costimulation molecules). Such inflammation can be induced in particular by infections with a virus showing a tropism for the target organ. Interestingly, more than one autoantigen will thus present increased immunogenicity, which fits with the multiplicity of autoreactive T cell specificities found in autoimmune diseases. Theiler's disease is probably the best illustration of this mechanism.[5]

2.2. Autoantigen mimicry

Activation of autoreactive T cells may also be achieved in the course of an immune response mounted against infectious agents whose antigens show cross reactivity with the target autoantigen. Rupture of tolerance to the common determinant is explained by carrier effect or epitope spreading. Rheumatic fever is best explained by this mechanism.

2.3. Superantigens

Microbial or endogenous retrovirus superantigens may provide activation of T cell subsets characterized by the expression of a given TCR Vβ chain. This activation which may ultimately be followed by apoptosis and deletion could explain the undesirable activation of autoreactive T cells included in this subset, the more so since some bias in T cell repertoire has been demonstrated in a number of autoimmune conditions.

2.4. Polyclonal B or T cell activation

The activation can even be more global if polyclonal B or T cell activation are operational. This is perhaps the case in non-organ specific autoimmune diseases, where endotoxins (LPS) may play such a role.

2.5. Autoantigen modification

It is well established that chemical modifications of an autoantigen (e.g. arsinilated thyroglobulin[6,7]) can exacerbate its capacity to induce an autoimmune disease. There is indication for the involvement of such a mechanism in various models of spontaneous thyroiditis whose incidence and severity are increased by addition of iodine in the diet which probably leads to thyroglobulin hyperiodination. At the cellular level, the mechanisms of rupture of ignorance is probably close to evoked for antigen mimicry.

3. T CELL-MEDIATED IMMUNOREGULATION MAY PREVENT OR DELAY AUTOIMMUNE DISEASES

Studies performed in NOD mice indicate that disease progression is slowed down by regulatory T cells. This concept is supported by several lines of experimental evidence that inactivation or depletion of selected T cell subsets accelerates disease onset.

Thymectomy at weaning accelerates diabetes onset.[8] This is not a general effect of thymectomy since no disease acceleration is seen when the thymus is removed at the adult age.

Cyclophosphamide treatment at high doses (150-200 mg/kg) triggers the onset of diabetes during the days following the drug administration.[9] The mechanism of action of cyclophosphamide is not established on a firm basis. It is most likely due to the acute depletion of some T cell subsets since diabetes onset can be prevented by injection of thymocytes or lymph node cells.[10] A direct toxic effect on β cells is unlikely since grafting of islets after the cyclophosphamide treatment does not prevent diabetes onset.[11] The deletion of regulatory T cells is also supported by the appearance of diabetogenic T cells (detected in transfer experiments) that are not found in matched non-cyclophosphamide treated NOD mice.[9]

Blockade of the B7-CD28/CTLA4 costimulatory pathway induces disease acceleration, also apparently through the depletion of a regulatory T cell subset. Such blockade can be obtained by inactivation of the CD28 gene[12] by transgenic overexpression of CTLA4-Ig[13] or by administration of anti-CTLA4[14] or anti anti-B7[12] antibodies.

Restriction of the T cell pool to T cell clones expressing the TCR of a diabetogenic CD4 T cell clone (after backcross with a CαKO mouse) also induces accelerated appearance of the disease (compared to the same transgenic mice on the NOD background) (L. Chatenoud, in preparation).

Similarly, induction of selective lymphocytopenia induces the spontaneous onset of autoimmune manifestations. Thus, thymectomy performed 2-3 days post natally in some mouse strains (e.g. BALB/c) induces a polyautoimmune syndrome.[15,16] Onset of this syndrome is prevented by injection of normal T cells arguing against the role of defective negative selection.

Adult thymectomy combined with irradiation induces the onset of diabetes and thyroiditis in Lewis rats, an effect also inhibited by normal T cells.[17]

One may mention in the same vein T cell mediated colitis[18] or polyautoimmune syndrome[19] induced by injecting immunodeficient mice (Nude or scid) with T cells depleted of selected T cell subset, respectively CD45RBlow or CD25+ T cells.[18,19]

All these experimental settings share the existence of a selective T cell lymphocytopenia. The fact that the autoimmune manifestations are prevented by the infusion of these lymphopenic animals with a T cell subset suggests that the T cell subset in question is endowed with regulatory properties.

Such lymphocytopenia-associated autoimmunity may also occur spontaneously as in the case of the BB rat in which autoimmune diabetes is preceded by a long phase of RT6+ T cell depletion.[4] The disease onset can be prevented by infusion of RT6+ T cells collected in non-diabetes prone littermates.[20]

4. VARIOUS TYPES OF REGULATORY T CELLS CAN BE EVIDENCED IN TRANSFER EXPERIMENTS

In all the lymphocytopenia-associated autoimmune states just described, as well as in several other models of autoimmunity not associated with lymphocytopenia (conventional

NOD mice, experimental allergic encephalomyelitis (EAE)) regulatory cells appeared to be of the CD4 phenotype. The detailed phenotype varies according to models, with the possibility that in most cases, only limited phenotyping was performed, precluding a comprehensive overview.

CD25 has been reported to be present on T cells preventing the onset of post-natal thymectomy.[15] It has recently been reported to be expressed by T cells preventing the effect of the B7-CD28 costimulatory pathway (Salomon and Bluestone, in preparation) and the CD45 RB cell depletion associated colitis (Powrie, in preparation). Data in conventional NOD mice are still unconclusive.

CD62L (l-selectin) is selectively expressed by thymocytes affording protection in the irradiated/thymectomised rat model[21] and in the NOD mouse.[22,23] In the latter model it has also been shown to be expressed in spleen cells from prediabetic mice and in cells mediating CD3 antibody mediated tolerance (Chatenoud, in preparation). Importantly, CD62L is not expressed in diabetogenic CD4 T cells.[24]

CD45. Several authors have attracted attention on the immunoregulatory role of CD45RBlow T cells. As mentioned above, Powrie's group demonstrated that CD45RBlow depleted normal T cells induced colitis when injected into athymic Nude mice.[18] Similarly, Shimada et al.[25] indicated that diabetes protective cells in young NOD mice were enriched in CD45RBlow cells. Lastly, Martins and Aguas[26] reported that the protective effect of Mycobacterium avium in NOD mice was associated with an increased number of T CD45RBlow cells.

CD38 is expressed by T cells affording protection in the lymphocytopenia associated colitis.[18] CD38$^+$ T cells show increased number in NOD mice protected from diabetes after treatment with *Mycobacterium avium*.[26]

5. AUTOANTIGEN SPECIFICITY OF REGULATORY T CELLS

There is only limited data on the specificity of autoimmune related T (ART) cells.

The most direct approach of such specificity was performed in the 80s by P. McCullagh first in the sheep, later in the rat. It was shown that removal of the thyroid rendered the sheep ultimately prone to reject the autologous thyroid in an organ specific fashion.[27] Similar data were obtained in rats by fetal destruction of thyrocytes by 131I treatment.[28-30] Direct demonstration of antigen-specific ART cells was not afforded but can be inferred from these data. These results have been confirmed by Mason's group.[30]

Other interesting data have recently been reported in orchiectomized and ovariectomized mice. Spleen cells from normal adult male mice were much more effective suppressors of autoimmune orchitis that follows day 3 thymectomy than spleen cells from female mice or male mice which had undergone a neonatal orchiectomy.[31] Additionally, Con A stimulated spleen cells from day 3 thymectomized male mice were less efficient at transferring oophoritis than spleen cells from female mice.[32] Thus male antigens in the first model and ovary antigens in the second one appear to be necessary to drive the corresponding autoimmune response following thymectomy. These results have been the matter of some controversy[33] which could be linked to the different number of transferred cells in each experiment. Note also that in the prostatitis model, spleen cells from males but not females or orchectomized males inhibited disease that follows day 3 thymectomy.[34] This effect was prostate-specific (without influence on sialoadenitis). Prostatitis developed after induction of prostate antigen following androgen treatment.

In the post natal thymectomy autoimmune syndrome there is also suggestion of an autoantigen specific regulation in spite of the multiplicity of the autoimmune traits induced by thymectomy. Indeed, when BALB/c mice overexpress in the thymus H$^+$/K$^+$ ATPase, the gastric target antigen in gastritis, they become specifically tolerant to that antigen and do not develop gastritis after thymectomy, while still showing all other manifestations of autoimmunity.[35]

No data is available in NOD mice. One may note, however, that ART cells that delay diabetes onset in BDC 2.5 transgenic mice are not any more detectable in BDC-Cα KO mice (Chatenoud, in preparation) suggesting that ART do not use the diabetogenic TCR.

6. ART CELL MODE OF ACTION

Most of the autoimmune models mentioned above are more likely mediated by Th1 cells. Subsequently, it has been assumed that ART cells that have been shown to be effective in these models could be of the Th2 type.

This view was initially supported by the observation that autoantigen-induced prevention of diabetes in NOD mice is associated with a Th2 polarization of the autoimmune response [cytokine production,[36,37] autoantibody isotypes[36-38]]. Anti-GAD autoreactive Th2 T cell clone was recently obtained that afford diabetes protection (Tisch, personal communication).

Much less evidence has been collected on ART cells that spontaneously occur in young animals independently of autoantigen external sensitization. Diabetes protection in NOD mice analysed in cotransfer experiments is not abrogated with anti-IL-4 and/or anti-IL-10 antibody treatment.[23] The post-natal thymectomy syndrome is still induced in IL-4 or IL-10 KO mice (Shevach, Ann. Rev. Immunol., in press). Consequently, one has to envision that AIRT cells do not act through the production of cytokines. It is intriguing to evoke the possibility that their function could involve a direct interaction with antigen presenting cells (APC) as has recently been suggested in several in vitro models of alloreactivity[39] or autoreactivity (W. van Eden, in preparation).

7. DIFFERENTIATION AND GENETIC CONTROL OF ART CELLS

If one favors the unicist hypothesis (putting together all ART cells described above), one may define some features of their differentiation :

differentiation takes place in the thymus (hence the acceleration of autoimmunity induced by thymectomy).[8,15]

- functional ART cells only appear in the periphery after 4-5 weeks (in NOD mice) whereas they are detected in the thymus as soon as 2 weeks of age (before ?).[40]

Little is known on the factors controlling ART cell differentiation. Particular interest has been recently given to the protective role of NKT cells. These T cells represent a newly described T cell population suspected to play a significant role in innate immunity and possibly in the differentiation of mainstream T cells, notably Th2 cells. They represent the major intrathymic source of IL-4 and NKT cell deprived mice ($\beta2m^{-/-}$) do not produce IgE after anti-IgD polyclonal stimulation.[41] Additionally, Vα14 TCR transgenic mice that overexpress NKT cells show a Th2 cell polarization with increased serum IgE levels.[42]

Their possible implication in predisposition to autoimmunity was initially suggested by their major deficiency noted in young NOD mice[43] as well as in SJL mice[44] known to develop chronic autoimmunity after sensitization with various autoantigens (e.g. MBP or autologous red blood cells). Importantly, this deficiency is both quantitative (number) and qualitative (IL-4 production). It is corrected by IL-7 treatment (in vitro or in vivo)[45] indicating that NKT cell precursors are not absent but do not mature normally.

The relationship between this NKT cell defect and autoimmunity predisposition has been further suggested by the partial prevention of diabetes observed in Vα14 transgenic

NOD mice[46] and the prevention of diabetes onset obtained by transfer of double negative (NKT cell enriched) thymocytes.[47] On the other hand, there is strong evidence to suggest that NKT cells are distinct from the ART cells described above. They are quasi absent from the thymus of 2 week-old NOD mice where ART cells are present in large amounts. IL-4 KO mice do not show accelerated diabetes onset[48] and NK T cells are $CD62L^-$,[21-23] whereas ART cells are $CD62L^+$.[22, 43]

An attractive hypothesis could implicate NK T cells in the control of the survival of ART cells, once they have seeded in the periphery.

At the genetic level, it is reasonable to assume that the genetic control of ART cell differentiation is related to the genetic predisposition to autoimmunity, particularly in cases where several autoimmune manifestations are noted in the same individual, as in the case of NOD mice which concomitantly show insulin-dependent diabetes mellitus, thyroiditis, sialitis, autoimmune hemolytic anemia and antinuclear antibodies.

Particular attention should be given to genes predisposing simultaneouly to these various autoimmune diseases as was recently described in our laboratory in NOD mice, a gene encoded on the 17th chromosome in an area close but distant from MHC (Garchon, in preparation). Similarly, one should pay particular attention to genes involved in the genetic predisposition to several autoimmune diseases with the exception of MHC genes which are putatively involved through the presentation of different autoantigens depending on the MHC alleles. A special note should be made in that regard on the autoimmune polyendocrine syndrome (APECED)[49] where a single gene mutation induces the onset of several autoimmune diseases. The nature of the cells expressing this gene (ART cells ?) will be important to determine.

8. ENVIRONMENT AND ART CELLS

Various environmental factors are known to trigger autoimmune diseases (as briefly discussed above). Conversely, other environmental factors appear to protect from autoimmunity. This is notably the case of infections. NOD mice or BB rats only develop a high incidence of diabetes when they are bred in specific pathogen free conditions.[50-52] Various infections have been shown to prevent disease onset. The protective role of infections could also be advocated at the origin of the North South gradient noted for a number of human autoimmune disorders.[53] It is tempting to believe that this protective role is mediated by ART cells. Indeed, diabetes protection in NOD mice afforded by administration of complete Freund's adjuvant or BCG can be transferred to untreated NOD mice by infusion of CD4 T cells from treated mice.[54]

It is interesting to note that the CFA effect may not mandatorily involve the production of IL-4 and IL-10 but rather that of IFNγ since it is still seen in IL-4 and IL-10 NOD KO mice but not in IFNγ KO mice (Rabinovitch, J. Immunol., in press).

9. RESTORATION OR INDUCTION OF ART CELLS IN AUTOIMMUNE STATES

It is not clear whether ART cell function declines when pathogenic autoimmunity appears or whether it is overriden by the intensity of the activation of autoreactive T cells. It remains identically interesting in any case to stimulate ART cells in autoimmunity prone mice. This can probably be achieved in various ways. We have been extensively involved in our laboratory in the stimulation of ART cells in NOD mice. Two approaches have been successfully used.

the protection thus afforded was mediated by CD4 CD62L[+] T cells. It could be broken by cyclophosphamide treatment.[55,56]

In the second approach, CD4[+] CD62L[+] T cells from overtly diabetic mice were transduced with the IL-4 gene using a retroviral vector. It was demonstrated in cotransfer experiments that their regulatory property was significantly enhanced (Yamamoto, in preparation).

Other methods of ART cell stimulation can be envisioned. The protective role of an agonistic anti-CD28 antibody probably relates to this mechanism.[57] A similar mechanism could be involved in the protective effect of non-depleting anti-CD4 antibodies demonstrated in several models.[58] Lastly, we have already mentioned the possible stimulation of ART cells afforded by various forms of immune stimulation (infections, administration of bacterial extracts,...). It will be important to determine at which stage of the disease these various maneuvers are efficacious.

Administration of autoantigens also stimulate regulatory T cells. One should, however, probably as discussed above distinguish these regulatory cells from ART cells discussed above. One may assume that Th2 cells are responsible for autoantigen-induced tolerance whereas there is no direct evidence for the Th2 nature of ART cells.

10. CONCLUSIONS

Data discussed above demonstrate the importance of regulatory T cells in the control of autoimmunity. One may schematically classify regulatory cells in three categories: Th2 cells, NKT cells and CD4 cells for which no evidence for Th2 phenotype has been brought. Table 1 presents the main characteristics of these three cell types. It is

Table 1. Regulatory lymphocytes involved in T-cell mediated diabetes prevention.

	Th2 cells	NK T cells	Spontaneous ART cells
Experimental situation	Administration of β-cell antigens CFA	Boosted by α-GalCer	Spontaneous disease Anti-CD3 treatment
Cytokine production	IL-4, IL-10	IL-4, IFN$_\gamma$	undetectable
Inhibition of IL-4 by IL-4/IL-10 antibody	+	+	-
Phenotype	CD4[+]	CD4/DN CD62L[-] CD25[-] CD122[+]	CD4[+] CD62L[+]
Antigen specificity	β-cell autoantigens Mycobacterial Antigens	Glycolipid (not β-cell related)	?
MHC restriction	Class II	Class I (CD 1d)	Class II (?)
TCR	?	Invariant α chain (Vα14-Jα281)	?

likely that the three cell types are physically distinct but one cannot exclude their filiation or the possibility that a given cell expresses different phenotypes according to immunologic situations. It is intriguing to observe that Th2 cells are essentially involved in experimentally-induced regulation (e.g. autoantigen-induced tolerance, CFA induced protection), whereas non-Th2 CD4 cells whose mode of action might not involve cytokines are essentially implicated in the control of spontaneously occurring autoimmunity (post-thymectomy autoimmune syndrome, NOD mouse). As far as NK T cells are concerned, it is difficult to determine whether they act in a direct fashion, or if they control the differentiation or the survival of the two other cell types or act pharmacologically by producing IL-4 (it is known that administration of IL-4 can control autoimmunity).

11. REFERENCES

1. Avrameas S, Ternynck T: The natural autoantibodies system: between hypotheses and facts. Mol Immunol 30:1133-1142, 1993
2. Shanmugam A, Copie-Bergman C, Hashim G, Rebibo D, Jais JP, Bach JF, Bach MA, Tournier-Lasserve E: Healthy monozygous twins do not recognize identical T cell epitopes on the myelin basic protein autoantigen. Eur J Immunol 24:2299-2303, 1994
3. Ohashi PS, Oehen S, Buerki K, Pircher H, Ohashi CT, Odermatt B, Malissen B, Zinkernagel RM, Hengartner H: Ablation of "tolerance" and induction of diabetes by virus infection in viral antigen transgenic mice. Cell 65:305-317,1991
4. Bach JF: Insulin-dependent diabetes mellitus as an autoimmune disease. Endocrine Rev 15:516-542,1994
5. Miller SD, Vanderlugt CL, Begolka WS, Pao W, Yauch RL, Neville KL, Katz Levy Y, Carrizosa A, Kim BS: Persistent infection with Theiler's virus leads to CNS autoimmunity via epitope spreading. Nat Med 3:1133-1136,1997
6. Weigle WO, Nakamura RM: Perpetuation of autoimmune thyroiditis and production of secondary renal lesions following periodic injections of aqueous preparations of altered thyroglobulin. Clin Exp Immunol 4:645-657,1969
7. Weigle WO: The production of thyroiditis and antibody following injection of unaltered thyroglobulin without adjuvant into rabbits previously stimulated with altered thyroglobulin. J Exp Med 122:1049-1062,1965
8. Dardenne M, Lepault F, Bendelac A, Bach JF: Acceleration of the onset of diabetes in NOD mice by thymectomy at weaning. Eur J Immunol 19:889-895,1989
9. Yasunami R, Bach JF: Anti-suppressor effect of cyclophosphamide on the development of spontaneous diabetes in NOD mice. Eur J Immunol 18:481-484,1988
10. Zhang ZL, Georgiou HM, Mandel TE: The effect of cyclophosphamide treatment on lymphocyte subsets in the nonobese diabetic mouse: a comparison of various lymphoid organs. Autoimmunity 15:1-10,1993
11. Charlton B, Bacelj A, Slattery RM, Mandel TE: Cyclophosphamide-induced diabetes in NOD/WEHI mice. Evidence for suppression in spontaneous autoimmune diabetes mellitus. Diabetes 38:441-447,1989
12. Lenschow DJ, Herold KC, Rhee L, Patel B, Koons A, Qin HY, Fuchs E, Singh B, Thompson CB, Bluestone JA: CD28/B7 regulation of Th1 and Th2 subsets in the development of autoimmune diabetes. Immunity 5:285-293,1996
13. Green JM, Noel PJ, Sperling AI, Walunas TL, Gray GS, Bluestone JA, Thompson CB: Absence of B7-dependent responses in CD28-deficient mice. Immunity 1:501-508,1994
14. Luhder F, Hoglund P, Allison JP, Benoist C, Mathis D: Cytotoxic T lymphocyte-associated antigen 4 (CTLA-4) regulates the unfolding of autoimmune diabetes. J Exp Med 187:427-432,1998

15. Nishizuka Y, Sakakura T: Thymus and reproduction: sex-linked dysgenesia of the gonad after neonatal thymectomy in mice. Science 166:753-755,1969

16. Sakaguchi S, Takahashi T, Nishizuka Y: Study on cellular events in postthymectomy autoimmune oophoritis in mice. I. Requirement of Lyt-1 effector cells for oocytes damage after adoptive transfer. J Exp Med 156:1565-1576,1982

17. Saoudi A, Seddon B, Fowell D, Mason D: The thymus contains a high frequency of cells that prevent autoimmune diabetes on transfer into prediabetic recipients. J Exp Med 184:2393-2398,1996

18. Groux H, Powrie F: Regulatory T cells and inflammatory bowel disease. Immunol Today 20:442-445,1999

19. Itoh M, Takahashi T, Sakaguchi N, Kuniyasu Y, Shimizu J, Otsuka F, Sakaguchi S: Thymus and autoimmunity: production of CD25+CD4+ naturally anergic and suppressive T cells as a key function of the thymus in maintaining immunologic self-tolerance. J Immunol 162:5317-5326,1999

20. Rossini AA, Faustman D, Woda BA, Like AA, Szymanski I, Mordes JP: Lymphocyte transfusions prevent diabetes in the Bio-Breeding/Worcester rat. J Clin Invest 74:39-46,1984

21. Seddon B, Saoudi A, Nicholson M, Mason D: CD4+CD8- thymocytes that express L-selectin protect rats from diabetes upon adoptive transfer. Eur J Immunol 26:2702-2708,1996

22. Herbelin A, Gombert JM, Lepault F, Bach JF, Chatenoud L: Mature mainstream TCR alpha beta(+)CD4(+) thymocytes expressing L-selectin mediate "active tolerance" in the nonobese diabetic mouse. J Immunol 161:2620-2628,1998

23. Lepault F, Gagnerault MC: Characterization of peripheral regulatory CD4(+) T cells that prevent diabetes onset in nonobese diabetic mice. J Immunol 164:240-247,1900

24. Lepault F, Gagnerault MC, Faveeuw C, Bazin H, Boitard C: Lack of L-selectin expression by cells transferring diabetes in NOD mice: insights into the mechanisms involved in diabetes prevention by Mel-14 antibody treatment. Eur J Immunol 25:1502-1507,1995

25. Shimada A, Rohane P, Fathman CG, Charlton B: Pathogenic and protective roles of CD45RB(low) CD4+ cells correlate with cytokine profiles in the spontaneously autoimmune diabetic mouse. Diabetes 45:71-78,1996

26. Martins TC, Aguas AP: A role for CD45RB(Low) CD38(1) T cells and costimulatory pathways of T-cell activation in protection of non-obese diabetic (NOD) mice from diabetes. Immunology 96:600-605,1999

27. King KJ, Hagan RP, Mieno M, McCullagh P: Cellular interactions during the development of autoimmunity in a fetal lamb model of self-antigen deprivation. Clin Immunol Immunopathol 88:56-64,1998

28. McCullagh P: Curtailment of autoimmunity following parabiosis with a normal partner. Immunology 71:595-597,1990

29. McCullagh P: The significance of immune suppression in normal self tolerance. Immunol Rev 149:127-153,1996

30. Seddon B, Mason D: Peripheral autoantigen induces regulatory T cells that prevent autoimmunity. J Exp Med 189:877-882,1999

31. Taguchi O, Nishizuka Y: Experimental autoimmune orchitis after neonatal thymectomy in the mouse. Clin Exp Immunol 46:425-434,1981

32. Taguchi O, Nishizuka Y: Self tolerance and localized autoimmunity. Mouse models of autoimmune disease that suggest tissue-specific suppressor T cells are involved in self tolerance. J Exp Med 165:146-156,1987

33. Smith H, Sakamoto Y, Kasai K, Tung KS: Effector and regulatory cells in autoimmune oophoritis elicited by neonatal thymectomy. J Immunol 147:2928-2933,1991

34. Taguchi O, Kontani K, Ikeda H, Kezuka T, Takeuchi M, Takahashi T, Takahashi T: Tissue-specific suppressor T cells involved in self-tolerance are activated extrathymically by self-antigens. Immunology 82:365-369,1994

35. Alderuccio F, Toh BH, Tan SS, Gleeson PA, Van Driel IR: An autoimmune disease with multiple molecular targets abrogated by the transgenic expression of a single autoantigen in the thymus. J Exp Med 178:419-426,1993

36. Tisch R, Liblau RS, Yang XD, Liblau P, McDevitt HO: Induction of GAD65-specific regulatory T-cells inhibits ongoing autoimmune diabetes in nonobese diabetic mice. Diabetes 47:894-899,1998

37. Elias D, Meilin A, Ablamunits V, Birk OS, Carmi P, Konen-Waisman S, Cohen IR: Hsp60 peptide therapy of NOD mouse diabetes induces a Th2 cytokine burst and downregulates autoimmunity to various beta-cell antigens. Diabetes 46:758-764,1997

38. Tian JD, Clare-Salzler M, Herschenfeld A, Middleton B, Newman D, Mueller R, Arita S, Evans C, Atkinson MA, Mullen Y, Sarvetnick N, Tobin AJ, Lehmann PV, Kaufman DL: Modulating autoimmune responses to GAD inhibits disease progression and prolongs islet graft survival in diabetes- prone mice. Nat Med 2:1348-1353,1996

39. Chai JG, Bartok I, Chandler P, Vendetti S, Antoniou A, Dyson J, Lechler R: Anergic T cells act as suppressor cells in vitro and in vivo. Eur J Immunol 29:686-692,1999

40. Boitard C, Yasunami R, Dardenne M, Bach JF: T cell-mediated inhibition of the transfer of autoimmune diabetes in NOD mice. J Exp Med 169:1669-1680,1989

41. Bendelac A, Rivera MN, Park SH, Roark JH: Mouse CD1-specific NK1 T cells: development, specificity, and function. Annu Rev Immunol 15:535-562,1997

42. Bendelac A, Hunziker RD, Lantz O: Increased interleukin 4 and immunoglobulin E production in transgenic mice overexpressing NK1 T cells. J Exp Med 184:1285-1293,1996

43. Gombert JM, Herbelin A, Tancrede-Bohin E, Dy M, Carnaud C, Bach JF: Early quantitative and functional deficiency of NK1(+)- like thymocytes in the NOD mouse. Eur J Immunol 26:2989-2998,1996

44. Yoshimoto T, Bendelac A, Hu-Li J, Paul WE: Defective IgE production by SJL mice is linked to the absence of CD4+, NK1.1+ T cells that promptly produce interleukin 4. Proc Natl Acad Sci USA 92:11931-11934,1995

45. Gombert JM, Tancrede-Bohin E, Hameg A, Leite-de-Moraes MC, Vicari A, Bach JF, Herbelin A: IL-7 reverses NK1+ T cell-defective IL-4 production in the non-obese diabetic mouse. Int Immunol 8:1751-1758,1996

46. Lehuen A, Lantz O, Beaudoin L, Laloux V, Carnaud C, Bendelac A, Bach JF, Monteiro RC: Overexpression of natural killer T cells protects V alpha 14-J alpha 281 transgenic nonobese diabetic mice against diabetes. J Exp Med 188:1831-1839,1998

47. Hammond KJ, Poulton LD, Palmisano LJ, Silveira PA, Godfrey DI, Baxter AG: alpha/beta-T cell receptor (TCR)(+)CD4(-)CD8(-) (NKT) thymocytes prevent insulin-dependent diabetes mellitus in nonobese diabetic (NOD)/Lt mice by the influence of interleukin (IL)-4 and/or IL-10. J Exp Med 187:1047-1056,1998

48. Wang B, Gonzalez A, Hoglund P, Katz JD, Benoist C, Mathis D: Interleukin-4 deficiency does not exacerbate disease in NOD mice. Diabetes 47:1207-1211,1998

49. Bjorses P, Aaltonen J, Horelli-Kuitunen N, Yaspo ML, Peltonen L: Gene defect behind APECED: a new clue to autoimmunity. Hum Mol Genet 7:1547-1553,1998

50. Ohsugi T, Kurosawa T: Increased incidence of diabetes mellitus in specific pathogen-eliminated offspring produced by embryo transfer in NOD mice with low incidence of the disease. Lab Anim Sci 44:386-388,1994

51. Hansen AK, Josefsen K, Pedersen C, Buschard K: Neonatal stimulation of beta-cells reduces the incidence and delays the onset of diabetes in a barrier-protected breeding colony of BB rats. Exp Clin Endocrinol 101:189-193,1993

52. Like AA, Guberski DL, Butler L: Influence of environmental viral agents on frequency and tempo of diabetes mellitus in BB/Wor rats. Diabetes 40:259-262,1991

53. Bach JF: Predictive medicine in autoimmune diseases: from the identification of genetic predisposition and environmental influence to precocious immunotherapy. Clin Immunol Immunopathol 72:156-161,1994

54. Qin HY, Sadelain MW, Hitchon C, Lauzon J, Singh B: Complete Freund's adjuvant-induced T cells prevent the development and adoptive transfer of diabetes in nonobese diabetic mice. J Immunol 150:2072-2080,1993

55. Chatenoud L, Thervet E, Primo J, Bach JF: Anti-CD3 antibody induces long-term remission of overt autoimmunity in nonobese diabetic mice. Proc Natl Acad Sci USA 91:123-127,1994

56. Chatenoud L, Primo J, Bach JF: CD3 antibody-induced dominant self tolerance in overtly diabetic NOD mice. J Immunol 158:2947-2954,1997

57. Arreaza GA, Cameron MJ, Jaramillo A, Gill BM, Hardy D, Laupland KB, Rapoport MJ, Zucker P, Chakrabarti S, Chensue SW, Qin HY, Singh B, Delovitch TL: Neonatal activation of CD28 signaling overcomes T cell anergy and prevents autoimmune diabetes by an IL-4-dependent mechanism. J Clin Invest 100:2243-2253,1997

58. Hayward AR, Schriber M, Cooke A, Waldmann H: Prevention of diabetes but not insulitis in NOD mice injected with antibody to CD4. J Autoimmun 6:301-310,1993

IMMUNE TOLERANCE AND THE NERVOUS SYSTEM

David E. Anderson, David A. Hafler

Center for Neurologic Diseases,
Brigham and Women's Hospital and Harvard Medical School
77 Avenue Louis Pasteur, Harvard Institutes of Medicine, Room 785
Boston, MA 02115
Tel: (617) 525-5330
Fax: (617) 525-5333
hafler@cnd.bwh.harvard.edu

1. INTRODUCTION

The immune system exists to protect the body from infection by a plethora of microorganisms, including bacteria, viruses, and parasites. This is accomplished by a variety of interdependent methods. The skin is an often overlooked but extremely effective barrier to infection, and is one of several innate mechanisms of immunity. NK cells, which appear to primarily recognize changes in the levels of major histocompatability complex class I (MHC cI) molecules expressed on cells within the body, represent another form of innate immunity. They represent a form of innate immunity because while they can protect the body against tumor cells or virally-infected cells, which often have altered levels of MHC cI expression, they have no "memory" of a given prior viral infection or particular type of tumor. This contrasts with T cells and B cells which comprise the specific immune response. A given T cell or B cell and its clonal progeny are all specific for a given foreign microbial antigen, and furthermore, can "remember" a prior encounter with that antigen and effectively remove it much more quickly upon secondary exposure to the antigen. This phenomena is the basis for vaccination. Thus, after vaccination with exposure to antigens from a particular virus or bacterium with adjuvant, there is activation and expansion of T cells and B cells with specific receptors for those particular antigens. If some years later the same individual is exposed to the infectious virus or bacteria, the T cells and B cells which have been previously activated and expanded will be mobilized very quickly and in most cases eliminate the infectious agent before it can do any harm to the body.

B cells and T cells respond to foreign antigens in fundamentally different ways. A B cell responds to a foreign antigen using a membrane-bound immunoglobulin (mIg) molecule as its receptor. This mIg receptor directly recognizes conformationally dependent stretches (epitopes) of a portion of foreign bacterial or viral protein. In marked contrast to B cells, T cells are critically dependent upon antigen presenting cells (APCs) such macrophages, dendritic cells, and B cells for their activation. T helper cells recognize short

Mechanisms of Lymphocyte Activation and Immune Regulation VIII
Edited by Sudhir Gupta, Kluwer Academic/Plenum Publishers, 2001

linear fragments of processed foreign antigens (peptides) presented on the surface of APCs by MHC class II (MHC cII) molecules.

While it is relatively easy for us to conceive of a protein fragment from a bacterium or virus as being foreign, it is not as readily apparent to the immune system. There is, after all, no inherent biochemical or structural difference between a protein fragment derived from a damaged cell within the body and a fragment of membrane from a virus. Yet the immune system must manage to reliably and consistently be in an activated state only in response to the viral protein fragment and not the fragment from its own tissues. This explains the necessity for immune tolerance, that is, the ability for the immune system to be tolerant of antigens from its own tissues yet respond to antigens from environmental sources, including bacteria, viruses, and parasites.

When an individual's immune system begins to attack its own tissue(s), we assume that immune tolerance has broken down and autoimmunity has developed. While there are many different autoimmune diseases, they do not all necessarily stem from similar defects in the maintenance of immune tolerance. Self antigens can induce tolerance by a variety of different mechanisms, and it is incumbent upon us to determine which mechanisms are defective in particular autoimmune diseases.

2. T CELL TOLERANCE IMPOSED BY THYMIC SELECTION

T cells must be able to recognize self MHC molecules presenting foreign antigens, yet during T cell selection within the thymus MHC molecules presenting predominantly self antigens are present to achieve this result. Two critical factors allow this goal to be achieved. First, T cell reactivity to foreign antigens is achieved by recognition of MHC molecules presenting cross-reactive self antigens, a process termed "positive selection," giving rise to thymocytes that are inherently autoreactive. Second, "negative selection" mediated by the same MHC complexes presenting self antigens eliminates all highly autoreactive thymocytes before they emigrate to the periphery as mature T cells. This process of negative selection is arguably the most important mechanism of ensuring immune tolerance. The following sections will first briefly describe the evidence for positive and negative T cell selection, and then focus on these two critical aspects of T cell selection, which are responsible for inducing central T cell tolerance.

2.1. Evidence for Positive and Negative Selection

The discovery of what has been termed "MHC restriction" was made by several groups throughout the 1970's and led to the award of two Nobel prizes. In the early 1970's Shreffler and Benaceraf determined that there was a requirement for a common MHC molecule for effective T cell:B cell interactions in adoptive transfer models of immunity in which B cells or T cells from one inbred strain of mice were transferred into another strain of irradiated recipient mice. In 1973 Shevach demonstrated that proliferative responses to antigen required MHC-matched APCs used to present antigen to the responding T cells. Using a newly developed *in vitro* cytotoxic assay for CD8+ T cells, Doherty and Zinkernagel similarly demonstrated MHC-restriction in the generation of anti-viral cytotoxic T lymphocyte (CTL) responses. Recent advances in technologies which allow the deletion of a particular gene (gene knockout mice) or which allow the constitutive expression of a gene of interest (transgenic mice) have allowed the role of the thymus in inducing central tolerance and imparting MHC restriction to be characterized more fully.

2.2. Anatomy and Location of T cell Selection

T cell specificity for self MHC molecules has been examined using a variety of approaches, including the use of MHC cI- and MHC cII-deficient mice and TCR transgenic mice in which the fate of the T cell repertoire with a single specificity can easily be evaluated. Bone marrow chimeric mice, in which bone marrow with a given MHC background from one mouse is used to reconstitute the immune system of a second strain of mice with a different, defined MHC specificity, have also been used extensively to elucidate the mechanisms of T cell positive and negative selection. Numerous investigations have collectively demonstrated that MHC restriction is determined by the MHC of the host thymus and is the result of positive selection. Taking advantage of MHC cII knockout mice and bone marrow chimeric mice expressing peptide/MHC cII complexes on either thymic stromal cells or on bone-marrow-derived hematopoetic cells, Glimcher and colleagues found that positive selection occurred only when peptide/MHC cII complexes were expressed by thymic stromal cells.[1] In an elegant follow-up study, this group used their peptide/MHC cII knockout mice and selectively reintroduced MHC cII complexes only on the thymic cortical epithelium (the medullary epithelium and bone marrow-derived cells were MHC II negative).[2] CD4+ cells were positively selected but no clonal deletion occurred. *In vitro* experiments demonstrated that cells from these mice proliferated vigorously to self APCs (were autoreactive), consistent with a lack of negative selection. Subsequent *in vivo* experiments demonstrated that adoptive transfer of these autoreactive T cells into irradiated syngeneic mice induced graft-versus-host disease (GVHD), and when transferred into non-irradiated syngeneic mice, induced hypergammaglobulinemia and autoantibody production.[3] These results convincingly demonstrate that positive selection in the absence of negative selection leads to generation of a polyclonal, pathogenic, autoreactive T cell repertoire.

While Glimcher and colleagues found that bone marrow-derived dendritic cells were not involved in positive selection, they did play a role in negative selection. In addition, several groups have confirmed that the thymic medullary epithelium mediates negative selection.[4, 5] These data further indicate that positive and negative selection occur in anatomically distinct sites.

2.3. The Cross-Reactive Basis of Selection

Allen and colleagues were the first to describe altered peptide ligands (APLs), which are peptides containing single amino acid substitutions which do not have appreciable effects on the binding of the peptides to the MHC molecule, but which do induce different signals in mature T cells through the TCR.[6] They have demonstrated the role of APLs in the selection of thymocytes,[7, 8] and one type of these APLs, antagonists, has been used by Hogquist and colleagues *in vitro* to encourage positive selection of thymocytes (described below). As mentioned earlier, the theories regarding the selection and imposed self-tolerance of peripheral mature T cells have implied that positive selection is based on extensive T cell cross-reactivity. For example, a thymocyte with low to moderate affinity for self peptide "A" might be positively selected and leave the thymus without being negatively selected because its TCR does not have too high an affinity for peptide "A" presented by MHC to warrant negative selection. In the periphery, this same mature T cell does not have any measurable reactivity to peptide "A" presented by MHC, but will be activated by foreign peptide "B" presented by the same MHC molecule. Thus, this thymocyte/T cell responds with different degrees of reactivity to two different peptides. It may be that for any given mature T cell, the self peptide used to select the cell as a thymocyte has very great similarity to the foreign peptide to which the mature T cell responds: the selecting peptide may be an antagonist peptide of the mature T cell differing from the agonist peptide used to activate the mature T cell by only one or two amino acid changes.

In some cases, however, the peptides used to select thymocytes may bear very little if any homology to the agonist foreign peptides used to activate the T cells within the periphery.[9] Using a novel technique, Kapler and Marrack's group created transgenic mice expressing a single peptide/MHC cII complex and found that a fairly diverse T cell repertoire was generated in these mice (about 20% the normal number of T cells).[10] This group subsequently demonstrated that T cells which were selected on a single peptide/MHC cII complex reacted with a diverse set of peptides bound by the same selecting MHC molecule, emphasizing the considerable degeneracy in peptide recognition during positive selection and peripheral T cell activation.[10] Nevertheless, while there is degeneracy in peptide recognition which can induce positive selection, one peptide cannot select the entire TCR repertoire.[11, 12]

2.4. Models of Positive and Negative Selection

Several models have been proposed to explain the mechanism by which a self-restricted, self-tolerant T cell repertoire is generated. One model suggested that perhaps the stage of thymocyte development at which point a given thymocyte interacted with a peptide/MHC complex dictated whether it was positively or negatively selected. Another model suggested that certain peptides might have an inherent ability to positively select while others served to negatively select thymocytes. Over time, however, these models have been disproved and will not be discussed further.

Currently, two primary models exist to explain T cell selection, the two models being slight conceptual variations of the same general mechanism. The differential avidity model of selection has been championed by the work of Ashton-Rickardt and Tonegawa. It postulates that the developmental fate of an immature thymocyte is dependent upon the avidity of interactions between thymocyte TCRs and peptide/MHC molecules on thymic stromal cells.[13, 14] This theory postulates that activating (or agonist) peptides induce positive selection of thymocytes when at low concentrations (low avidity), while at higher doses the same peptide encourages negative selection high avidity). In the efficacy model of thymocyte selection advocated by Hogquist and Bevan, the quality of the signal a peptide/MHC complex induces in a thymocyte, in addition to the number of peptide/MHC complexes present, determines the outcome of the selection process for a given thymocyte. This group has used variants of peptides for which a given thymocyte is specific, called antagonists, which induce signals which inhibit rather than activate mature T cells expressing the same TCR found within the periphery. When these antagonist peptides were used in *in vitro* assays of thymocyte selection, this group found that while agonist peptides induced negative selection at all concentrations, including very low concentrations at which the Ashton-Rickardt group saw positive selection of thymocytes, the antagonist peptides were found to positively select thymocytes when present at concentrations above a certain threshold.[15, 16] A schematic incorporating the salient features of these two models is presented in Figure 1.

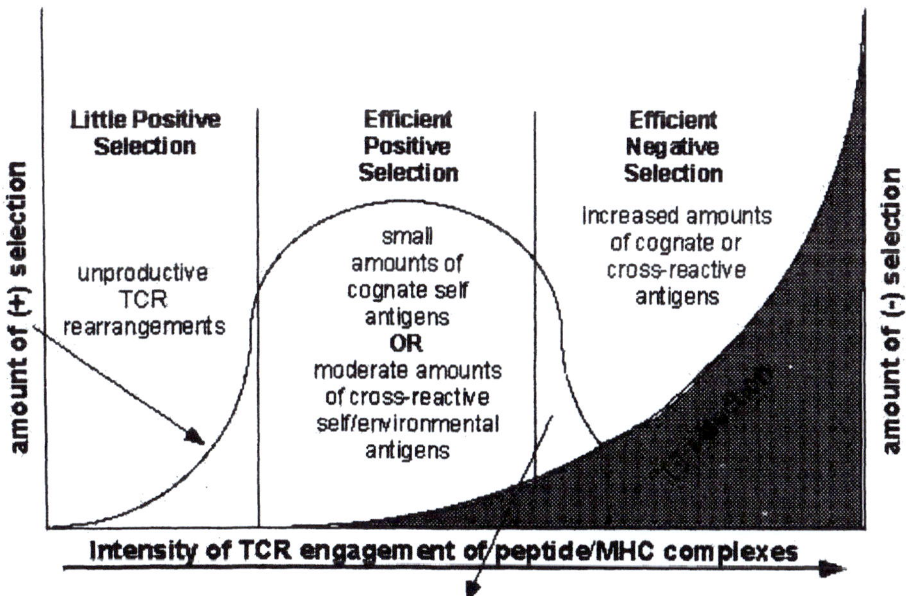

Figure 1. A small number of thymocytes are positively selected yet escape negative selection (central tolerance). Factors which allow for escape from negative selection include a lack or insufficient amount of autoantigen in the thymus or inefficient binding and presentation of autoantigen by the MHC molecules.

Several other groups have provided data to support the notion that the avidity of interaction between a thymocyte and selecting peptide/MHC complexes dictates the outcome of the selection process. Sebzda et al.[17] used the same TCR transgenic mice as Tonegawa's group, but used β2-microglobulin knockout mice instead of TAP knockout mice to eliminate contaminating endogenous peptide presentation by MHC molecules. Using only a peptide agonist (a defined viral peptide recognized by the TCR), they found that low concentrations mediated positive selection whereas high concentrations led to negative selection. Cook et al.[18] have shown that rather than modulating the amount of peptide used to select a thymocyte with a TCR specific for the peptide/MHC complex, they could modulate the level of the MHC molecules and achieve a similar effect. Thus, TCR transgenic mice were bred to express different levels of MHC molecules on the thymic stroma; at low levels of selecting peptide/MHC complexes the thymocytes were positively selected whereas mice bred to express high levels of the selecting peptide/MHC molecules were found to negatively select the vast majority of thymocytes. In another study, surface plasmon resonance was used to directly measure the kinetics of TCR interactions with peptides known to positively and negative select thymocytes.[19] The results suggested that the affinity of the peptide for the TCR correlated with the outcome of selection, such that positive selection occurred over a 1-log range of peptide starting 3-fold below the affinity required for negative selection. However, the affinity of the TCR for peptide/MHC complex was not an absolute predictor of the outcome of selection, and the authors suggested that clustering of co-receptors might play a role in the selection process by influencing the overall avidity of a T cell for APC. The presence or absence of various co-receptors and adhesion molecules might just as easily affect the quality of the signals directed through the TCR.

Two additional models have been postulated to explain T cell selection and the induction of central T cell tolerance. A "gemisch" of peptides may be responsible for positive selection of thymocytes: the selection of any one TCR may not be dependent on one self peptide, rather, the summation of each peptide/MHC complex of very low affinity may add up to high ligand density and positive selection. In this way, none of the cooperating, selecting ligands would be individually detected as agonists or antagonists.[9] Alternately, Matzinger has suggested that while thymocytes require negative selection in order to prevent widespread autoimmunity mediated by recent thymic emigrants, thymocytes may not go through any positive selection process. Instead, peripheral tissue damage due to physical trauma or insult or microbial infection could induce some kind of "danger" signal to the immune system encouraging the activation of T cells.[20] Little direct evidence, however, has as yet been demonstrated to support these alternate models of T cell selection.

2.5. Role of Apoptosis in Negative Selection

In 1972 Kerr, Currie, and Wyllie proposed the term apoptosis to describe a common series of morphologic changes that accompanied the death of cells from a wide variety of tissues. Apoptosis is fundamentally different from cell necrosis in many important ways. It requires new gene expression and is triggered by the appearance or loss of an external signal which leads to the activation of an internal cell death program. In contrast, cell necrosis does not appear to require the expression of new mRNAs or proteins and is believed to be initiated by cellular damage that disrupts osmotic balance. Perhaps most importantly, as a cell undergoes apoptosis it may break up into apoptotic bodies, but these are sealed and maintain their osmotic gradients; there is no spilling of intracellular contents, and no provocation of inflammation.[21]

During T cell selection, apoptosis serves to eliminate precursor cells with non-rearranged or aberrantly rearranged non-functional receptors (those thymocytes which are not positively selected) and is also responsible for the deletion of autoreactive T cells in the thymus due to negative selection. Using three-color immunofluorescence and FACS analysis on frozen human thymic tissue and freshly isolated human thymocytes, Le et al.[22] documented the role of apoptosis in central tolerance. They found that the majority of apoptotic thymocytes were localized to the cortical-medullary junction, where anatomically negative selection is known to occur.

2.6. Central Tolerance and Autoimmunity

Based on the fundamental investigations into how central tolerance functions to prevent the development of autoreactive T cells, many ideas have emerged to explain autoimmune diseases. Wraith and colleagues have theorized that low affinity interactions between autoantigens and MHC molecules may undermine the efficacy of their presentation in the thymus, thus enabling autoreactive T cells to escape self-tolerance. To test this theory his group used a TCR transgenic mice specific for the central nervous system (CNS) autoantigen myelin basic protein (MBP) pl-11 and APLs based on this MBP epitope which had greater binding affinity for MHC cII molecules. Consistent with their theory, they found that the *in vivo* administration of the APLs resulted in deletion of T cells while administration of native peptide had no effect.[23] Other groups have suggested similar explanations for the tendency of certain strains of mice to be more susceptible to certain autoimmune diseases, such as diabetes in NOD mice.[24, 25] It is clear that autoreactive cells can be found in the peripheral blood of both normal individuals and patients with autoimmune diseases such as multiple sclerosis (MS).[26] Thus, the presence of autoreactive cells in the periphery alone is an insufficient explanation for the development of autoimmunity. Either the pool of autoreactive cells is much larger in patients with autoimmune diseases, they are in a different functional state,[27-29] they have different effector profiles, or they have different thresholds of activation

3. PERIPHERAL MECHANISMS OF T CELL TOLERANCE

3.1. T cell Anergy

Efficient T cell activation is dependent upon two signals: an antigen-specific signal delivered through the TCR and a second costimulatory signal which functions to induce the secretion of T cell growth factors such as IL-2. The best characterized costimulatory pathway is the B7 pathway.[30, 31] Two related B7 costimulatory molecules, B7-l and B7-2, are expressed on APCs, although with different kinetics and expression patterns. B7-2 is found on most APCs at low but constitutive levels whereas B7-1 is generally absent until an APC becomes activated, at which time it upregulates the expression of both molecules. These molecules direct signals into T cells through two receptors, CD28 and CTLA4. It is clear that signals directed through CD28 enhance T cell activation, while signals delivered through CTLA4 serve to attenuate T cell activation.[32, 33]

The consequence of T cell activation with peptide/MHC cII complexes in the absence of B7 costimulation was first reported by Jenkins and Schwartz.[34] They found that when ECDI-treated splenocytes (which effectively fixes the cell surface and inactivates many surface molecules) were used *in vitro* as APCs, they failed to stimulate proliferation by antigen-specific normal T cell clones and instead induced a state of long-term unresponsiveness termed anergy. This T cell unresponsiveness was also induced *in vivo* by the i.v. administration of antigen-coupled splenocytes prepared by ECDI treatment. The results were not due to extensive MHC cII complex alteration, as anti-MHC cII mAbs prevented this anergy induction, suggesting that antigen presentation was taking place and needed for the anergy induction. The authors proposed that the ECDI treatment impaired an additional APC signal necessary to induce IL-2 production and T cell proliferation. The anergic state did not seem to involve inhibition of the IL-2 receptor pathway, however, as T cell clones unable to respond to Ag/MHC restimulation responded normally to exogenous IL-2.

Extensive research has been conducted by a multitude of groups trying to induce anergy in the hopes of preventing rejection in various models of transplantation and ameliorating a wide range of autoimmune diseases. Miller and colleagues have used a system very similar to the one used by Jenkins and Schwartz to demonstrate that chemical cross-linking of APCs pulsed with a variety of autoantigens and viral antigens known to induce various forms of experimental autoimmune encephalomyelitis (EAE), a rodent model of MS, can induce tolerance and ameliorate/delay EAE.[35-37] Working with human T cell clones, Boussiotis et al.[38] found that stimulation of T cell clones with fibroblasts expressing only MHC cII molecules (no B7-1 or B7-2 molecules present) resulted in anergy induction. If unmanipulated APCs were incubated with CTLA4Ig and mixed with T cells an anergic state was similarly induced. In this case the CTLA4Ig, a chimeric molecule with the binding domain of the high-affinity B7 receptor CTLA4 fused to the constant portion of an immunoglobulin heavy chain, present in the cultures binds to the B7 molecules present on APCs and prevents the delivery of a costimulatory signal to the responding T cells in the culture.

The immunosuppressive effects of *in vivo* CTLA4Ig administration have been well documented.[39-41] It has been shown to induce T cell tolerance and inhibit both primary and secondary T cell responses, as well as inhibit antibody production. Furthermore, it has been used in several models of autoimmune disease. It was used in the treatment of mice with an experimental form of lupus and found to block autoantibody production and prolong life, even when treatment was delayed until the most advanced stage of clinical illness.[42] Arima et al.[43] examined the effects of CTLA4Ig administration on the induction of EAE in Lewis rats. CTLA4Ig administration 8 times before/immediately after immunization with spinal cord homogenate was able to prevent the development of EAE, and this was reversed by the administration of rIL-2 (suggesting that indeed, the cells were rendered anergic). However, administration of CTLA4Ig twice after immunization with spinal cord homogenate slightly enhanced the severity of disease, suggesting that blocking the CD28 pathway is most useful before T cells have been activated and/or clonally expanded.

CTLA4Ig administration has shown considerable promise in transplantation as well. Bluestone and colleagues have demonstrated that CTLA4Ig administration to mice receiving human pancreatic islets induced long-term, donor-specific tolerance (graft acceptance greater than 45 days relative to 5 days for controls).[44] While CTLA4Ig treatment alone has shown promise in transplantation in the treatment of autoimmune diseases, a combination of CTLA4Ig and blocking mAbs directed against CD40 or CD40L have shown considerably greater *in vivo* effects. This is presumably due in part to the fact that CD40 engagement on APCs induces the upregulation of the B7 costimulatory molecules. One group demonstrated convincingly that while administration of CTLA4Ig or anti-CD 40 mAb could inhibit alloreactive mixed-lymphocyte-reactions (MLRs) *in vitro* and prevent skin graft rejection *in vivo* to some extent alone, they were extremely potent when combined together.[45] Another group similarly compared the administration of anti-human CD40L mAb and CTLA4Ig alone or together in their ability to prevent renal allograft rejection in rhesus macaques. The allografts were mismatched for both MHC cI and MHC cII antigens. Administration of both anti-CD40L mAb and CTLA4Ig was more effective that either agent alone, and there was no generalized immunosuppression in that *in vitro* MLR responses were normal. Two monkeys receiving the optimal regimen were healthy for over 150 days.[46] Finally, Sarvetnick's group has shown that anti-CD40L mAbs given at 3 weeks of age to NOD mice, which develop spontaneous diabetes at about 10 weeks of age, prevented insulitis and diabetes. When measured *in vitro*, the mAb treatment reduced antigen-specific T cell proliferation and IFNγ production.[47]

3.2. Activation-Induced Peripheral Cell Death

Lenardo and colleagues were the first to characterize what has been termed activation-induced cell death (AICD).[48] They noticed that suppressed T cell proliferation *in vitro* at high antigen doses correlated with high IL-2 production and apoptosis. To explore this observation further, T cells were stimulated with antigen and then placed in various concentrations of IL-2 for 2 days, and subsequently restimulated with antigen. They found that apoptosis occurred to degrees correlating with the amount of IL-2 present before antigen restimulation. Using TCR transgenic mice specific for an encephalitogenic MBP peptide, this group found that repeated immunizations of soluble MBP or MBP peptide could delete up to 80% of the MBP-specific T cells. The authors argued that T cells "sensed" the intensity of an immune response by the level of cell cycling after initial antigen stimulation, and that further antigen stimulation attenuated the immune response by decreasing the number of responding cells and in turn downmodulating the amount of IL-2 production.

Using IL-2 receptor α-chain (CD25) knockout mice crossed with TCR transgenic mice, Van Parijs et al.[49] found that, as expected, these cells proliferated and survived less well than wild type T cells, and did not differentiate into effector cells. Stimulation in the presence of IL-15 compensated for this defect to some extent, likely because it signals through a common IL-2 receptor γ chain. Most importantly, these mice did not undergo AICD; IL-15 could not compensate for IL-2 and did not therefore increase the degree of AICD in the CD25 knockout TCR transgenic mice. These results confirmed *in vivo* that IL-2 responsiveness is necessary for AICD.

Many *in vitro* and *in vivo* models of AICD have found that signaling through CD28 on T cells can provide protection against apoptosis.[50-53] This protection has been found to correlate with an increase in the intracellular expression of the anti-apoptotic molecule bcl-xl. Because it is clear that IL-2 is required both to initiate/sustain an immune response, and to help terminate a T cell immune response via AICD, the role of costimulation in AICD could change depending upon to what extent activation versus apoptosis was being favored by the IL-2 induced by the costimulatory signal. Thus, CD28 costimulation likely has a role not only in initiating a primary response, but also in maintaining T cell viability during a primary response, and finally in terminating excess antigen-specific T cells at the conclusion of an immune response.

One mechanism by which IL-2 may encourage AICD of T cells is through the upregulation of FasL. In IL-2 receptor γ-chain knockout mice there is a defect in peripheral CD4+ T cell deletion.[54] Interestingly, there is also a lack of FasL expression upon activation, presumably explaining the lack of peripheral apoptosis. Thus, the data suggest that there is a requirement for the common γ-chain to convey the IL-2 signal into the cell for upregulation of FasL and subsequent AICD. Indeed, several reports appeared at the same time demonstrating that the interaction between Fas and FasL on single T cell clones or hybridomas could induce apoptosis after TCR triggering and T cell activation.[55-57] Furthermore, Abbas and Marshak-Rothstein have demonstrated that mice with functional defects in the expression of Fas or FasL fail to undergo AICD and succumb to massive lymphoproliferation and autoimmune disease.[58] The kinetics of FasL expression are consistent with the notion that it plays a critical role in AICD because it is almost exclusively expressed by activated T cells.[59] Immune privileged sites, such as the eyes or testes, may indeed be protected by FasL expression on tissues within these sites.[60,61] In an elegant study, Griffith et al.[62] used normal and gld mice (FasL-deficient) and showed that viral infection resulted in T cell inflammation in the eyes of gld mice but not in normal mice due to apoptosis in normal mice. Furthermore, Fas^+ but not Fas^- tumor cells were killed when placed in the anterior chamber of normal mice, further emphasizing that FasL expression may contribute to immune privilege.

Marrack and colleagues have studied the mechanism of superantigen-mediated AICD *in vivo*. In mice injected with superantigen, T cells activated by the superantigen proliferate extensively, then rapidly die. The authors found that LPS administration was capable of preventing superantigen-driven T cell death and that TNFα mediated part of the effect.[63]. Another group elucidated the mechanism by which LPS protected $CD4^-$ T cells from superantigen-mediated AICD. A small number of T cells were transferred into a syngeneic host and then the fate of the cells was tracked after immunization of the mice with the antigen for which the cells were specific. Co-injection of LPS significantly enhanced the clonal expansion of the transferred cells, their migration in germinal centers, and the help they provided for antibody production. Most importantly, the proinflammatory cytokines IL-1 and TNFα could substitute for LPS, suggesting that the adjuvant effects of LPS are mediated in large part by proinflammatory cytokines.[64]

3.3. T cell Suppression by Cytokines

CD4+ T helper cells can be broadly categorized into one of several subsets, based on the cytokines produced by the T cells after activation.[65,66] T helper 1 (Th1) cells secrete proinflammatory cytokines such as IFNγ, TNFα and lymphotoxin (LT), which enhance APC activation and the clearance of many intracellular pathogens, whereas Th2 cells secrete cytokines such as IL-4, IL-5, and IL-13. These cytokines aid in antibody class-switching and elimination of many blood-borne infectious agents.[65] The cytokines produced by each T helper cell subset can themselves negatively regulate the differentiation of the other subset. Thus, IFNγ produced by Th1 cells inhibits the differentiation of naïve T cells into Th2 cells and IL-4 secreted by Th2 cells can inhibit the differentiation of Th1 cells. The cytokines IL-10 and TGFβ, secreted by T regulatory 1 (Tr1) and Th3 cells, respectively, are very potent suppressors of T cell activation. IL-10 has been shown to be very potent at inhibiting LPS-induced macrophage cytokine production[67, 68] and a combination of IL-4 and IL-10 were very effective at inhibiting a DTH reaction *in vivo*.[69] *In vivo* administration of rhIL-10 to humans has further demonstrated the ability of IL-10 to inhibit proinflammatory cytokine production.[70] TGFβ has been shown to inhibit T cell proliferation by a variety of mechanisms.[71] The strategy of encouraging the development of Th2, Tr1, or Th3 T cells in an effort to inhibit pathogenic Th1 cells has been termed immune deviation.

Oral tolerance is currently being tested as a way to suppress the immune response in patients with autoimmune diseases such as MS, rheumatoid arthritis (RA), and diabetes. Weiner and Hafler have conducted numerous studies demonstrating that the primary effect of oral tolerance is the expansion of antigen-specific T cells which secrete TGFβ.[72, 73] In one study, Weiner and colleagues examined brain tissue for cytokine expression in lesions from Lewis rats at the peak of EAE.[74] Th1 cytokines were present during peak disease, but were diminished when rats were fed MBP. In addition, TGFβ was present. In a phase II clinical study of the effects of oral administration of bovine MBP to patients with the relapsing-remitting form of MS, T cell lines generated from patients fed oral MBP were found to secrete significantly more TGFF in response to MBP than T cell lines from patients not fed oral MBP.[75]

While inducing TGFβ secretion from autoreactive T cells by the oral administration of autoantigens is one form of inducing immune suppression, many studies have demonstrated that shifting autoreactive T cells towards a Th2 phenotype shows promise in the prevention of Th1-mediated autoimmune diseases as well.[76] In one study neuroantigen-specific T cells were differentiated *in vitro* into Th1 or Th2 phenotypes then adoptively transferred into syngeneic mice and assessed for their ability to induce EAE.[77] The data indicated that Th1 but not Th2 cells consistently induced disease, but that Th2 cells administered together with Th1 cells could not prevent disease. However, another study demonstrated that the adoptive transfer of Th2 MBP-specific T cells into syngeneic mice could reduce the incidence and eliminate the relapse of EAE.[78] Another elegant study confirmed these results.[79] EAE-susceptible SJL mice were immunized with the protein KLH in incomplete freunds adjuvant (IFA) to induce Th2 cells reactive to KLH. EAE was then induced in the same mice using whole guinea pig myelin in complete freunds adjuvant (CFA), which is known to induce proinflammatory cytokines and induce potent disease. The authors found that if they included KLH with the whole myelin/CFA immunogen in mice which had been preimmunized with the KLH in IFA, that the severity of disease was prevented or greatly reduced. These results indicated that the cytokine microenvironment present at the time autoreactive T cells are primed/activated could influence their pathogenicity. Moreover, Th2 cytokines appeared to protect against disease, and this influence was effective even if cells were primed systemically rather than within the CNS. Consistent with this, Racke et al.[80] demonstrated that *in vivo* administration of IL-4 to mice ameliorated EAE clinical disease, induced Th2 MBP-reactive T cells, diminished demyelination, and reduced inflammatory cytokines.

Two recent papers suggest future directions in attempts to modulate Th1-mediated autoimmune diseases by cytokine-mediated suppression. Shaw et al.[81] described an attempt to ameliorate EAE induced in an adoptive transfer system by transducing a MBP-specific T cell hybridoma with IL-4 and adoptively transferring it to mice. The authors demonstrated a beneficial effect of local delivery of IL-4 in the CNS during the induction of disease and emphasized the promise of local delivery of IL-4 into the CNS. A similar paper appeared at the same time using a proteolipid protein (PLP)-specific T cell clone transfected with IL-10.[82] The clone could significantly inhibit EAE disease induction/severity and to some extent ameliorate disease after onset.

3.4. T cell Ignorance and Immune Privileged Sites

Several sites within the body have been designated by some investigators as "immune privileged" sites because of the supposed relative difficulty of B cell and T cell migration into the sites. The CNS has been touted in the past as a prime example, particularly given the tight endothelial junctions which help form the blood-brain barrier. Experimental data generated in recent years has clearly shown, however, that T cell migration into such sites occurs routinely and can do so with little pathogenic consequence.[83]

There are likely several reasons that the presence of autoreactive T cells within an organ expressing the autoantigen may not necessitate an autoimmune response. As described earlier, expression of FasL by the target tissue may induce apoptosis in autoreactive T cells. Alternately, the T cells may be rendered anergic. Indeed, a paper from Goodnow and colleagues suggests that multiple mechanisms of tolerance exist, and that self-reactive T cells do not necessarily cause tissue-destruction when they encounter self antigen within an organ, a phenomena called clonal ignorance.[84] In this study the protein HEL was expressed on the thyroid epithelium, pancreatic islet cells, or systemically in the presence of TCR transgenic T cells specific for HEL. They found that the mechanisms used to induce tolerance depended on the location and/or concentration of antigen. Thus, when the self antigen was expressed in either the thyroid or pancreas, T cells were less tolerant and thyroiditis or insulitis was present, although there was no destruction of self tissue; there was almost total deletion of T cells when HEL was expressed systemically. Other groups have seen similar influences of the site and concentration of antigen exposure to autoreactive T cells on the tolerance induced.

Hammerling's group generated transgenic mice which expressed a foreign non-self MHC cI molecule on hepatocytes.[85] Low levels of the MHC cI molecule were detected on mouse hepatocytes, but high levels could be induced by LPS treatment. The group found that rather than simply ignoring the foreign tissue-specific antigen, T cells actually downregulated their expression of antigen-specific TCRs and were tolerant. When MHC cI expression was increased via LPS treatment, a portion of the previously tolerant cells appeared to be deleted. They concluded that the dose of antigen influences the degree of tolerance induced, and that tolerant T cells are not necessarily refractory to further contact with the tolerogen and even greater tolerance induction (and/or deletion). Tolerance, they hypothesize, might be the result of multiple interactions with antigen. Similarly, in another system TCR transgenic T cells specific for male antigen were injected into thymectomized mice with various ratios of male and female bone marrow cells.[86] The authors found that when antigen persisted, anergy was induced with high concentrations of antigen while exhaustion occurred at lower antigen concentrations.

In summary, the vast majority of autoreactive T cells are deleted within the thymus by MHC complexes presenting a myriad of self antigens. This form of T cell tolerance is called central tolerance. However, some presumably lower-affinity autoreactive T cells do escape central tolerance and exist within the periphery of healthy, normal individuals without causing any pathology due to many peripheral mechanisms of tolerance. The many mechanisms of T cell tolerance are summarized in Table 1.

Table 1. Mechanisms of T cell tolerance.

Type of Tolerance	Mechanism	Relevant Molecules/Ligands
CENTRAL	Deletion via Apoptosis	MHC molecules presenting (self) autoantigens
PERIPHERAL	Anergy	B7-1/2:CD28 and CD40:CD40L pathways
	Activation-induced Cell Death (AICD)	IL-2, FasL
	Immune Deviation/ T cell suppression	Anti inflammatory cytokines (IL-4 IL-10, TGFβ)
	Immune Privilege	FasL

4. B CELL TOLERANCE

Activation of B cells and antibody isotype class switching are both critically dependent upon T cell help delivered by surface costimulatory molecules, such as CD40L expressed by activated T cells, or in the form of cytokines secreted by Th2 cells, such IL-4 and IL5. Because of this dependence on T cells for complete maturation and activation, T cell tolerance indirectly implies B cell tolerance. However, several mechanisms do exist to ensure autoreactive B cells are not given the chance to receive aberrant T cell help. These mechanisms are presented in Table 2 and discussed below.

Table 2. Mechanisms of B cell tolerance.

Type of Tolerance	Mechanism	Relevant Molecules/Ligands
CENTRAL	Receptor Editing	Early B cell engagement of self antigen
	Clonal Deletion	Late B cell engagement of self antigen
PERIPHERAL	Lack of T cell help	CD40:CD40L pathway and cytokines (IL-4, 5)
	Deletion	Membrane-bound self antigen
	Anergy	Soluble self antigen

4.1. B Cell Affinity Maturation

Within the paracortex of lymph nodes (or an analogous structure in the spleen) resting T cells can be found as well as specialized APCs (follicular dendritic cells) which are presumably responsible for the initial stimulation of the T cells.[87] Once stimulated, these T cells move to the paracortex/primary follicular junctions which are dominated by naïve B cells which express mostly surface IgM or IgD receptors. In response to cytokines secreted by antigen-stimulated T cells nearby, a few antigen-specific B cell clones will divide to form germinal centers, made up mainly of these B cell blasts and a few activated CD4+ T cells. Activated T cells with similar antigen specificity can then encourage the survival and expansion of B cells. Within these germinal centers, follicular dendritic cells capture and present intact antigen to B cells. In addition, several days after an immune response has begun, antigen-antibody complexes may also bind to FDCs. It is at this point that somatic mutations occur, which ultimately leads to the generation of higher affinity germinal center B cells. This process, called "affinity maturation," is based on a competition among the different B cell clones with similar specificity. Those which after somatic mutations have a higher affinity for the antigen with which they interact will compete for the antigen more effectively and continue their antigen-dependent clonal expansion. This is analogous to the process of positive selection of T cells, such that only cells which can productively engage antigen are given the chance to mature into competent lymphocytes of the periphery.

4.2. Clonal vs. Receptor Selection

In an elegant series of recent papers, Nemazee's group has demonstrated that two different mechanisms of immune tolerance are utilized within the bone marrow when immature B cells encounter self antigen, and that the mechanism employed depends on the maturation state of the B cell. The first mechanism used to impart B cell tolerance is referred to as receptor editing. If an immature self-reactive B cell encounters self antigen in the bone marrow, this process promotes rearrangement of the second Ig receptor light chain in the hopes of altering the BCR specificity. In one study, two-thirds of autoreactive immature B cells were found to undergo receptor editing without any significant apoptosis.[88] Another study found that later in B cell development after receptor editing has occurred, immature B cell engagement with self antigen leads to apoptosis within the bone marrow.[89] Constitutive transgenic expression of the anti-apoptotic molecule bcl-2 could not prevent central tolerance impart by self antigen expression within the bone marrow, but it did enhance the receptor editing process.[90]

4.3. B cell Clonal Deletion vs. Anergy

The bone marrow is the primary lymphoid organ responsible for development of B cells while the secondary lymphoid organs, such as the spleen and lymph nodes, are specialized for the generation of immune responses by B cells upon antigen recognition. The lymph nodes drain most intercellular spaces via afferent lymphatic flow, while Peyer's Patches (PP), the appendix, and mesenteric nodes drain the intestines. Foreign antigens that enter the blood are cleared by the spleen.

The clever use of transgenic mice by both Nemazee and Goodnow has allowed several insights to be gained into the various mechanisms by which B cell tolerance is induced. Nemazee and Burki[91] generated transgenic mice expressing anti MHC mIg receptors which produced high levels of anti-MHC cI antibodies detectable in their sera.

However, when they crossed them to mice expressing the same MHC cI autoantigen, the B cells were deleted. This mechanism of tolerance involving the deletion of autoreactive B cell clones differed from another form of B cell tolerance first described by Nossal and Pike.[92] These authors described the presence of B cells with autoantigen-specific Ig receptors which were functionally incapable of responding to antigen; this form of tolerance was referred to as anergy. Goodnow and colleagues confirmed both of these results using a similar system but with an elegant twist. They generated several different lines of transgenic mice. They generated transgenic mice which expressed a protein, Hen Egg Lysozyme (HEL) in either a membrane or soluble form, as well as a B cell transgenic mice in which all the membrane Ig receptors were specific for the HEL antigen. When the BCR transgenic mice were crossed to the two different types of HEL transgenic mice, the autoreactive immature B cells were eliminated in the bone marrow upon recognition of membrane-bound antigens, confirming the results of Nemazee and colleagues. However, in mice in which the autoantigen was secreted into the serum, the B cells were not deleted, rather, they were anergized.[93, 94] When anergized, the B cells were found to downregulate the expression of their surface Ig receptors and they had a greatly shortened lifespan of just 3-4 days. Thus, the work of these groups has demonstrated that B cell tolerance can be induced in different ways depending on the form in which antigen is recognized by an autoreactive B cell.

Using these same double transgenic systems, the above groups discovered an additional mechanism of ensuring B cell tolerance. When anergic B cells were transferred into mice with a diverse repertoire of B cell receptors (e.g. non-Ig receptor transgenic mice) but which expressed soluble HEL protein systemically (which ensured the induction of anergy in HEL-specific B cells), Cyster et al.[95] found that these anergic B cells had an altered pattern of migration from the bone marrow. After leaving the bone marrow and being rendered anergic by engagement of soluble HEL in the periphery, the anergic B cell eventually migrated into the spleen. Within the spleen, however, these cells moved only to the outer part of the T cell area, the peri-arteriolar lymphocyte sheath (PALS). Because they were anergic, they could not receive antigenic or costimulatory signals from the few T cells present in this area, signals which would encourage the B cells to proceed into the follicles where extensive T cell help and B cell somatic hypermutation occurs.

5. IMMUNE TOLERANCE AND AUTOIMMUNE DISEASES

As described in detail above, there are many mechanisms by which the immune system seeks to ensure immune tolerance. Central tolerance of both T cells and B cells involves primarily a deletional mechanism mediated by apoptosis. However, it is clear that autoimmune T cells and B cells do exist in the periphery of normal individuals and experimental animals, and yet no autoimmune disease arises. Thus, peripheral mechanisms of tolerance play a significant role in preventing the development of autoimmunity as well. One of the biggest challenges to immunologists studying autoimmune diseases is to determine which mechanisms of ensuring tolerance have failed so that we can utilize the appropriate treatment for the disease.

For example, suppose that a failure to delete autoreactive T cells within the thymus proves to be responsible for a given autoimmune disease. If so, therapies are currently available and actively being refined which can delete autoreactive T cells within the thymus. Bone marrow chimerism, in which an allogeneic source of bone marrow from a healthy donor is transplanted into a diseased individual, has been shown to induce very efficient deletion of autoreactive cells.[96] Indeed, bone marrow transplantation is currently being explored for the treatment of several autoimmune diseases including MS and RA.[97-99]

If, however, there are defects in peripheral tolerance mechanisms, treatments using CTLA4Ig and anti-CD40L mAbs may be capable of inducing anergic states in autoreactive T cells. Alternately, oral tolerance might be used to suppress the autoreactive response or shift it towards a more benign type of response.

5.1. EAE-Autoimmune Model of Multiple Sclerosis

Experimental Autoimmune Encephalomyelitits or EAE is one animal model for MS, and can be induced by injecting animals with either myelin basic protein (MBP), or proteolipid apoprotein (PLP) in adjuvant. MBP and PLP are the two most abundant myelin proteins in the CNS and are the only two proteins discovered to date that induce EAE. EAE can be induced in virtually all species and has been most extensively studied in the mouse and rat. The pathologic picture of EAE, especially in chronic relapsing forms, can be very similar to MS. EAE is mediated by T cells that react to brain proteins. MBP or PLP specific CD4+ T cell clones can be isolated from animals with EAE and cause relapsing EAE when injected into normal animals. Transfer of serum from animals with EAE does not cause disease, although injections of antibodies against myelin components such as myelin
oligodendrocyte glycoprotein (MOO) can exacerbate EAE and increase demyelination.

It has been argued for years as to whether EAE is a good model for MS and the answer will probably not be known until the cause of MS is defined. Nonetheless the biologic principles learned from the study of EAE can be generalized to other experimental autoimmune process such as experimental autoimmune neuritis (EAN, induced with the injection of adjuvant with the peripheral myelin protein P0) or experimental autoimmune myasthenia gravis (EAMG, induced by injecting adjuvant with acetylcholine receptor(AchR)). Though myasthenia gravis and EAMG are both mediated by T cells, the effector mechanism of the human and experimental disease are both mediated by anti-AchR antibodies.

5.2. Virus Induced Models of T cell Mediated CNS Inflammation

A second model of T cell mediated CNS white matter disease is induced by Theiler's virus, a picornavirus. In the Theiler's virus mouse model of demyelination, the virus infecting the brain serves as the target of an immune attack. Demyelination is dependent upon an intact T cell mediated immune response. In the Theiler's virus model, T cells reacting to myelin antigens are not found. This is a model of direct viral induced inflammatory demyelination. Another model of virus induced autoimmune disease is induced by coronavirus infection in young rat brain. Here, viral infection of the brain leads to activation of MBP reactive T cells that are capable of transferring disease to naïve animals not previously infected with virus. The coronavirus model of demyelination illustrates how CNS viral infection may induce autoreactive T cells. It has been postulated that HTLV-I associated myelopathy may have a similar etiology to Theiler's virus encephalitis where HTLV-I infection of the CNS would be followed by immune recognition and secondary CNS inflammation mediated by viral reactive T cells.

6. REFERENCES

1. Markowitz, J., H. J. Auchincloss, M. Gursby, et al. 1993. Class II-positive hematopoietic cells cannot mediate positive selection of CD4+ T lymphocytes in class II-deficient mice. *Proc. Natl. Acad. Sci. USA* 90:2779.

2. Laufer, T., J. DeKoning, J. Markowitz, et al. 1996. Unopposed positive selection and autoreactivity in mice expressing class II MHC only on thymic cortex. *Nature* 383:81.

3. Lauter, T.M., L. Fan, and L.H. Glimcher. 1999. Self-reactive T cells selected on thymic cortical epithelium are polyclonal and are pathogenic *in vivo. J. Immunol.* 162:5078.

4. Surh, C. and J. Sprent. 1994. T-cell apoptosis detected in situ during positive and negative selection in the thymus. *Nature* 372:100.

5. Oukka, M., E. Colucci-Guyon, P. Tran et al. 1996. CD4 T cell tolerance to nuclear proteins induced by medullary thymic epithelium. *Immunity* 4:545.

6. Evavold, B. D. and P. M. Allen. 1991. Separation of IL-4 production from Th cell proliferation by an altered T cell receptor ligand. *Science* 252:1308.

7. Allen, P. 1994. Peptides in positive and negative selection: a delicate balance. *Cell* 76:593.

8. Vidal, K., B. Hsu, C. Williams et al. 1996. Endogenous altered peptide ligands can affect peripheral T cell responses. *J. Exp. Med.* 183:1311.

9. Bevan, M. 1997. In thymic selection, peptide diversity gives and takes away. *Immunity* 7:175.

10. Ignatowicz, L., J. Kappler and P. Marrack. 1996. The repertoire of T cells shaped by a single MHC/peptide ligand. *Cell* 84:521.

11. Grubin, C., S. Kovats, P. deRoos et al. 1997. Deficient positive selection of CD4 T cells in mice displaying altered repertoires of MHC class II-bound self-peptides. *Immunity* 7:197.

12. Surh, C., D. Lee, W. Fung-Leung et al. 1997. Thymic selection by a single MHC/peptide ligand produces a semidiverse repertoire of CD4+ T cells. *Immunity* 7:209.

13. Ashton-Rickardt, P. G., A. Bandeira, J. R. Delaney et al. 1994. Evidence for a differential avidity model of T cell selection in the thymus. *Cell* 76:651.

14. Girao, C., Q. Hu, J. Sun et al. 1997. Limits to the differential avidity model of T cell selection in the thymus. *J. Immunol.* 159:4205.

15. Hogquist, K. A., S. C. Jameson, W. R. Health et al. 1994. T cell receptor antagonist peptides induce positive selection. *Cell* 76:17.

16. Hogquist, K., S. Jameson and M. Bevan. 1995. Strong agonist ligands for the T cell receptor do not mediate positive selection of functional CD8+ T cells. *Immunity* 3:79.

17. Sebzda, E., V. Wallace, J. Mayer et al. 1994. Positive and negative thymocyte selection induced by different concentrations of a single peptide. *Science* 263:1615.

18. Cook, J., E. Wormstall, T. Hornell et al. 1997. Quantitation of the cell surface level of Ld resulting in positive versus negative selection of the 2C transgenic T cell receptor *in vivo. Immunity* 7:233.

19. Alam, S., P. Travers, J. Wung et al. 1996. T-cell-receptor affinity and thymocyte positive selection. *Nature* 381:616.

20. Langman, R. and M. Cohn. 1996. Terra Firma: A retreat from danger. *J. Immunol.* 157:4273.

21. Cohen, J. 1993. Apoptosis. *Immunol. Today* 14:126.

22. Le, P., H. Maecher and J. Cook. 1995. In situ detection and characterization of apototic thymocytes in human thymus. Expression of bcl-2 *in vivo* does not prevent apoptosis. *J. Immunol.* 154:4371.

23. Liu, G., P. Farichild, R. Smith et al. 1995. Low avidity recognition of self-antigen by T cells permits escape from central tolerance. *Immunity* 3:407.

24. Kanagawa, O., S. Martin, B. Vaupel et al. 1998. Autoreactivity of T cells from nonobese diabetic mice: an I-Ag7-dependent reaction. *Proc. Natl. Acad. Sci. USA* 95:1721.

25. Ridgway, W.M., Fasso, M. and Fathman, C.G. 1999. A new look at MHC and autoimmune disease. *Science* 284:749.

26. Ota, K., M. Matsui, E. Milford et al. 1990. T-cell recognition of an immunodominant myelin basic protein epitope in multiple sclerosis. *Nature* 346: 183.

27. Scholz, C., K. Patton, D. Anderson et al. 1998. Expansion of autoreactive T cells in multiple sclerosis is independent of exogenous B7 costimulation. *J. Immunol.* 160:1532.

28. Zhang, J., S. Markovic, J. Raus et al. 1994. Increased frequency of IL-2 responsive T cells specific for myelin basic protein and proteolipid protein in peripheral blood and cerebrospinal fluid of patients with multiple sclerosis. *J. Exp. Med.* 179:973.

29. Lovett-Racke, A., J. Trotter, J. Lauber

30. et al. 1998. Decreased dependence of myelin basic protein-reactive T cells on CD28-mediated costimulation in multiple sclerosis patients. *J. Clin. Invest.* 101 :725.

31. Lenschow, D. and J.A. Bluestone. 1996. CD28/B7 system of T cell costimulation. *Ann. Rev. Immunol.* 14:233.

32. Tivol, E., A.N. Schweitzer, and A.H. Sharpe. 1996. Costimulation and autoimmunity. *Curr. Opin. Immunol.* 822.

33. Linsley, P. 1995. Distinct roles for CD28 and cytotoxic T lymphocyte-associated molecule-4 receptors during T cell activation? *J. Exp. Med.* 182:289.

34. Thompson, C.B., and Allison, J.P. 1997. The emerging role of CTLA-4 as an immune attenuator. *Immunity* 7:445.

35. Jenkins, M. K., D. M. Pardoll, J. Mizuguchi et al. 1987. T-cell unresponsiveness *in vivo* and *in vitro*: fine specificity of induction and molecular characterization of the unresponsive state. *Immunological Reviews* 95:113.

36. Karpus, W., J. Pope, J. Peterson et al. 1995. Inhibition of Theiler's virus-mediated demyelination by peripheral immune tolerance induction. *J. Immunol.* 155:947.

37. Kennedy, K., W. Smith, S. Miller et al. 1997. Induction of antigen-specific tolerance for the treatment of ongoing, relapsing autoimmune encephalomyelitis: a comparison between oral and peripheral tolerance. *J. Immunol.* 159:1036.

38. Vandenbark, A., B. Celnik, M. Vainiene et al. 1995. Myelin antigen-coupled splenocytes suppress experimental autoimmune encephalomyelitis in Lewis rats through a partially reversible energy mechanism. *J. Immunol.* 155:5861.

39. Boussiotis, V., D. Barber, T. Nakaria et al. 1994. Prevention of T cell energy by signaling through the gamma c chain of the IL-2 receptor. *Science* 266:1039.

40. Linsley, P. S., P. M. Wallace, J. Johnson et al. 1992. Immunosuppression *in vivo* by a soluble form of the CTLA-4 T cell activation molecule. *Science* 257:792.

41. Milich, D., M. Chen, J. Hughes et al. 1998. The secreted hepatitis B precore antigen can modulate the immune response to the nucleocapsid: a mechanism for persistence. *J. Immunol.* 160:2013.

42. Wallace, P., J. Rodgers, G. Leytze et al. 1995. Induction and reversal of long-lived specific unresponsiveness to a T-dependent antigen following CTLA4Ig treatment. *J. Immunol.* 154:5885.

43. Finck, B., P. Linsley and D. Wofsy. 1994. Treatment of murine lupus with CTLA4Ig. *Science* 265:1225.

44. Arima, T., A. Rehman, W. Hickey et al. 1996. Inhibition by CTLA4Ig of experimental allergic encephalomyelitis. *J. Immunol.* 156:4916.

45. Lenschow, D. J., Y. Zeng, J. R. Thistlethwaite et al. 1992. Long-term survival of xenogeneic pancreatic islet grafts induced by CTLA4Ig. *Science* 257:789.

46. Larsen, C., E. Elwood, D. Alexander et al. 1996. Long-term acceptance of skin and cardiac allografts after blocking CD40 and CD28 pathways. *Nature* 381:434.

47. Kirk, A., D. Harlan, N. Armstrong et al. 1997. CTLA4-Ig and anti-CD40 ligand prevent renal allograft rejection in primates. *Proc. Natl. Acad. Sci USA* 94:8789.

48. Balasa, B., T. Krahl, G. Patstone et al. (1997). CD40 ligand-CD40 interactions are necessary for the initiation of insulitis and diabetes in nonobese diabetic mice. *J. Immunol.* 159:4620.

49. Critchfield, J., M. Racke, J. Zuniga-Pflucker et al. 1994. T cell deletion in high antigen dose therapy of autoimmune encephalomyelitis. *Science* 263: 1139.

50. Van Parijs, L., A. Biuckians, A. Ibragimov et al. 1997. Functional responses and apoptosis of CD25 (IL-2R alpha)-deficient T cells expressing a transgenic antigen receptor. *J. Immunol.* 158:3738.

51. Noel, P., L. Biose and C. Thompson. 1996. CD28 costimulation prevents cell death during primary T cell activation. *J. Immunol.* 157:636.

52. Boise, L., A. Minn, P. Noel et al. 1995. CD28 costimulation can promote T cell survival by enhancing the expression of Bcl-XL. *Immunity* 3:87.

53. Mueller, D., S. Seiffert, W. Fang et al. 1996. Differential regulation of bcl-2 and bcl-x by CD3, CD28, and the IL-2 receptor in cloned CD4+ helper T cells. A model for the long-term survival of memory cells. *J. Immunol.* 156:1764.

54. Radvanyi, L., Y. Shi, H. Vaziri et al. 1996. CD28 costimulation inhibits TCR-induced apoptosis during a primary T cell response. *J. Immunol.* 156:1788.

55. Nakajima, H. and W. Leonard. 1997. Impaired peripheral deletion of activated T cells in mice lacking the common cytokine receptor gamma-chain: defective Fas ligand expression in gamma-chain-deficient mice. *J. Immunol.* 159:4737.

56. Brunner, T., R. Mogil, D. LaFace et al. 1995. Cell-autonomous Fas (CD95) /Fas ligand interaction mediates activation-induced apoptosis in T-cell hybridomas. *Nature* 373:441.

57. Dhein, J., H. Walczak, C. Baumler et al. 1995. Autocrine T-cell suicide mediated by APO-1/(Fas/CD95). *Nature* 373:438.

58. Ju, S., D. Panka, H. Cut et al. 1995. Fas (CD95)/FasL interactions required for programmed cell death after T-cell activation. *Nature* 373:444.

59. Ettinger, R., J. Wang, P. Bossu et al. 1994. Functional distinctions between MRL-lpr and MRL-gld lymphocytes. Normal cells reverse the gld but not lpr immunoregulatory defect. *J. Immunol.* 152:1557.

60. Nagata, S. and P. Golstein. 1995. The Fas death factor. *Science* 267:1449.

61. Streilein, J. 1996. Unraveling immune privilege. *Science* 270:1158.

62. Streilein, J., B. Ksander and A. Taylor. 1997. Immune deviation in relation to ocular immune privilege. *J. Immunol.* 158:3557.

63. Griffith, T., T. Brunner, S. Fletcher et al. 1996. Fas ligand-induced apoptosis as a mechanism of immune privilege. *Science* 270:1189.

64. Vella, A., J. McCormack, P. Linsley et al. 1995. Lipopolysaccharide interferes with the induction of peripheral T cell death. *Immunity* 2:261.

65. Pape, K., A. Khoruts, A. Mondino et al. 1997. Inflammatory cytokines enhance the *in vivo* clonal expansion and differentiation of antigen-activated CD4+ T cells. *J. Immunol.* 159:591.

66. Abbas, A., K. Murphy and A. Sher. 1996. Functional diversity of helper T lymphocytes. *Nature* 383:787.

67. O'Garra, A. (1998) Cytokines induce the development of functionally heterogeneous T helper cell subsets. *Immunity* 8, 275-283.

68. Fiorentino, D., A. Zlotnik, T. Mosmann et al. 1991. IL-10 inhibits cytokine production by activated macrophages. *J. Immunol.* 147:3815.

69. Wang, P., P. Wu, M. Siegel et al. 1994. IL-10 inhibits transcription of cytokine genes in human peripheral blood mononuclear cells. *J. Immunol.* 153:811.

70. Powrie, F., S. Menon and R. Coffman. 1993. Interleukin-4 and interleukin-10 synergize to inhibit cell-mediated immunity *in vivo.* Eur. *J. Immunol.* 23:3043.

71. Pajkrt, D., L. Camoglio, M. Tiel-van Buul et al. 1997. Attenuation of proinflammatory response by recombinant human IL-10 in human endotoxemia: effect of timing of recombinant human IL-10 administration. *J. Immunol.* 158:3971.

72. Bright, J., L. Kerr and S. Sriram. 1997. TGF-beta inhibits IL-2-induced tyrosine phosphorylation and activation of Jak-1 and Stat 5 in T lymphocytes. *J. Immunol.* 159:175.

73. Chen, Y., J. Inobe and H. Weiner. 1995. Induction of oral tolerance to myelin basic protein in CD8-depleted mice: both CD4+ and CD8+ cells mediated active suppression. *J. Immunol.* 155:910.

74. Weiner, H. 1997. Oral tolerance: immune mechanisms and treatment of autoimmune diseases. *Immunol. Today* 18:335.

75. Khoury, S. J., W. W. Hancock and H. L. Weiner. 1992. Oral tolerance to myelin basic protein and natural recovery from experimental autoimmune encephalomyelitis are associated with down-regulation of inflammatory cytokines and differential upregulation of TGF-D, IL-4 and PGE expression in the brain. *J. Exp. Med.* 176:1355.

76. Fukaura, H., S. Kent, M. Pietrusewicz et al. 1996. Induction of circulating myelin basic protein and proteolipid protein-specific transforming growth factor-01-secreting Th3 T cells by oral administration of myelin in multiple sclerosis. *J. Clin. Invest.* 98:70.

77. Nicholson, L. and V. Kuchroo. 1996. Manipulation of the Thl/Th2 balance in autoimmune disease. *Curr. Opin. Immunol.* 837.

78. Khoruts, A., S. Miller and M. Jenkins. 1995. Neuroantigen-specific Th2 cells are inefficient suppressors of experimental autoimmune encephalomyelitis induced by effector Thl cells. *J. Immunol.* 155:5011.

79. Cua, D., D. Hinton and S. Stohlman. 1995. Self-antigen-induced Th2 responses in experimental allergic encephalomyelitis (EAE)-resistant mice. Th2-mediated suppression of autoimmune disease. *J. Immunol.* 155:4052.

80. Falcone, M. and B. Bloom. 1997. A T helper cell 2 (Th2) immune response against non-self antigens modifies the cytokine profile of autoimmune T cells and protects against experimental allergic encephalomyelitis. *J. Exp. Med.* 185:901.

81. Racke, M., A. Bonomo, D. Scott et al. 1994. Cytokine-induced immune deviation as a therapy for inflammatory autoimmune disease. *J. Exp. Med.* 180:1961.

82. Shaw, M., J. Lorans, A. Dhawan et al. 1997. Local delivery of interleukin 4 by retrovirus-transduced T lymphocytes ameliorates experimental autoimmune encephalomyelitis. *J. Exp. Med.* 185:1711.

83. Mathisen, P., M. Yu, J. Johnson et al. 1997. Treatment of experimental autoimmune encephalomyelitis with genetically modified memory T cells. *J. Exp. Med.* 186:159.

84. Williams, K. and W. Hickey. 1995. Traffic of hematogenous cells through the central nervous system. *Curr. Top. Microbiol. Immunol.* 202:221.

85. Akkaraju, S., W. Ho, D. Leong et al. 1997. A range of CD4 T cell tolerance: partial inactivation to organ-specific antigen allows nondestructive thyroiditis or insulitis. *Immunity* 7:255.

86. Ferber, I., G. Schenrich, J. Schenkel et al. 1994. Levels of peripheral T cell tolerance induced by different doses of tolerogen. *Science* 263:674.

87. Rocha, B., A. Grandien and A. Freitas. 1995. Anergy and exhaustion are independent mechanisms of peripheral T cell tolerance. *J. Exp. Med.* 181:993.

88. Weissman, I. 1994. Developmental switches in the immune system. *Cell* 76:207.

89. Melamed, D. and D. Nemazee. 1997. Self-antigen does not accelerate immature B cell apoptosis, but stimulates receptor editing as a consequence of developmental arrest. *Proc. Natl. Acad. Sci. USA* 94:9267.

90. Melamed, D., R. Benschop, J. Cambier et al. 1998. Developmental regulation of B lymphocyte immune tolerance compartmentalized clonal selection from receptor selection. *Cell* 92:173.

91. Lang, J., B. Arnold, G. Hammerling et al. 1997. Enforced Bc1-2 expression inhibits antigen-mediated clonal elimination of peripheral B cells in an antigen dose-dependent manner and promotes receptor editing in autoreactive, immature B cells. *J. Exp. Med.* 186:1513.

92. Nemazzee, D. and K. Burki. 1989. Clonal deletion of B lymphocytes in a transgenic mouse bearing anti-MHC class I antibody genes. *Nature* 337:562.

93. Nossal, G. and B. Pike. 1980. Clonal energy: persistence in tolerant mice of antigen-binding B lymphocytes incapable of responding to antigen or mitogen. *Proc. Natl. Acad. Sci. USA* 77:1602.

94. Goodnow, C. C. 1992. Transgenic mice and analysis of B-cell tolerance. *Annual Review of Immunology* 10:489.

95. Fulcher, D. and A. Basten. 1994. Reduced life span of anergic self-reactive B cells in a double-transgenic model. *J. Exp. Med.* 179:125.

96. Cyster, J., S. Hartley and C. Goodnow. 1994. Competition for follicular niches excludes self-reactive cells from the recirculating B-cell repertoire. *Nature* 371:389.

97. Nikolic, B. and M. Sykes. 1997. Bone marrow chimerism and transplantation tolerance. *Curr. Opin. Immunol.* 9:634.

98. Burt, R., W. Burns and S. Miller. 1997. Bone marrow transplantation for multiple sclerosis: returning to Pandora's box. *Immunol. Today* 18:559.

99. Fassas, A., A. Anagnostopoulos, A. Kazis et al. 1997. Peripheral blood stem cell transplantation in the treatment of progressive multiple sclerosis: first results of a pilot study. *Bone Marrow Trans.* 20:631.

100. Krance, R. and M. Brenner. 1998. BMT beats autoimmune disease. *Nature Med.* 4:153.

FUNCTIONAL ROLE OF EPITOPE SPREADING IN THE CHRONIC PATHOGENESIS OF AUTOIMMUNE AND VIRUS-INDUCED DEMYELINATING DISEASES

Stephen D. Miller and Todd N. Eagar

Department of Microbiology-Immunology and
  Interdepartmental Immunobiology Center
Northwestern University Medical School
303 E. Chicago Avenue
Chicago, IL 60611

INTRODUCTION

Precise delineation of the mechanisms involved in the initiation and progression of autoimmune diseases is one of the most enigmatic issues in the field of immunology. It is clear that the etiology of many autoimmune diseases is influenced by both genetic and environmental factors. This is certainly true for multiple sclerosis (MS), a T cell-mediated demyelinating disease of the central nervous system (CNS) characterized by autoimmune responses to myelin proteins such as myelin basic protein (MBP), proteolipid protein (PLP), and myelin oligodendrocyte glycoprotein (MOG) (1-4). Epidemiological evidence strongly supports the hypothesis that many forms of human MS are initiated by a viral infection (5). The caveat of this hypothesis, which is strongly supported by epidemiological evidence, is that no particular virus or viruses have been consistently isolated from MS lesions. Thus, exploring the mechanisms by which infectious agents trigger autoimmune diseases is of great interest. Three mechanisms have been suggested to explain how a viral infection could lead to autoimmunity: viral *superantigens* that non-specifically lead to activation of autoreactive T cells (6); *molecular mimicry* between the pathogen and self antigens which leads to direct activation of T cells induced against epitopes on the pathogen that are cross reactive with self epitopes (7); and *epitope spreading* evoked by virus-specific T cells that result in bystander damage to self tissue with consequent autoantigen release (8) or direct virus-induced release of self antigens (9), resulting in *de novo* activation of autoreactive T cells.

Our laboratory has been studying the pathogenesis and immunoregulation of two different SJL mouse models of MS over the past several decades. The first paradigm, relapsing experimental autoimmune encephalomyelitis (R-EAE) models the relapsing-remitting form of human MS and is induced by active immunization with myelin proteins such as PLP, MBP or MOG, or by the transfer of activated cells specific for the immunodominant encephalitogenic epitopes encoded within the myelin proteins. This disease is characterized by a moderate to severe acute paralytic phase followed by remission and subsequent relapses (10). The second paradigm, Theiler's murine encephalomyelitis virus (TMEV)-induced demyelinating disease (TMEV-IDD) models the chronic-progressive form of human MS and is induced by intracerebral infection with Theiler's virus, a picornavirus and a natural mouse pathogen (11-15). This disease is characterized by a slow progressive development of spastic paralytic symptoms which first onset approximately 30 days post-infection. Experiments from our laboratory have demonstrated that the chronic stages of both R-EAE and TMEV-IDD are associated with the development of T cell responses to endogenous myelin peptides recruited secondary to acute CNS damage, a process termed epitope spreading (8,16,17). This review will summarize evidence showing that T cells specific for endogenous myelin epitopes play a significant pathologic role in tissue damage during the relapsing clinical episodes of R-EAE and during the chronic-progressive stages of TMEV-IDD. In addition, evidence showing that antigen presenting cells within the CNS acquire the ability to present endogenous myelin epitopes, thus providing a mechanistic basis to explain epitope spreading, will be summarized.

FUNCTIONAL ROLE OF EPITOPE SPREADING IN MEDIATING RELAPSES IN PEPTIDE-INDUCED R-EAE

We have shown that in R-EAE induced by the immunodominant PLP139-151 epitope, recovery from the acute clinical episode is accompanied by the expansion of PLP178-191-specific T cells (*i.e.*, intramolecular epitope spreading) in the peripheral lymphoid tissues. SJL mice remitting from acute R-EAE induced with the weakly encephalitogenic MBP84-104 epitope develop PLP139-151-specific T cell responses concomitant with disease relapse (*i.e.*, intermolecular epitope spreading) (17). This pattern of intra- and intermolecular epitope spreading is depicted in Figure 1. Interestingly, the pattern of epitope spreading follows a hierarchical order associated with the relative encephalitogenic dominance of the myelin epitopes (PLP139-151 > PLP178-191 > MBP84-104).

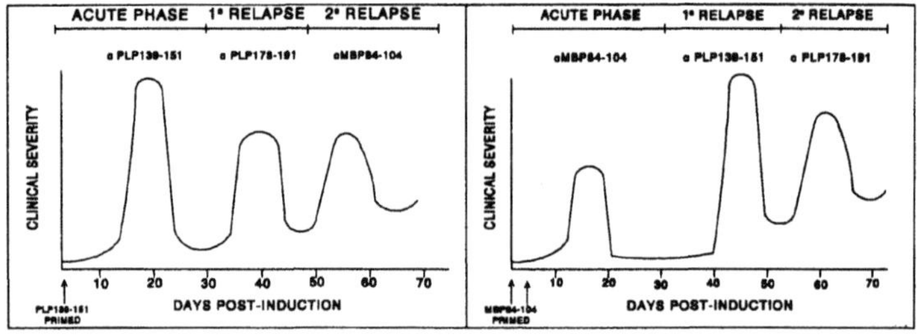

Figure 1. Pattern of intra- and intermolecular epitope spreading in R-EAE induced by priming of SJL mice with PLP139-151 (Left Panel) and MBP84-104 (Right Panel).

Diversification of autoimmune responses thus appears to be a common sequelae to myelin destruction in murine models of EAE and, importantly, recent evidence suggests that a similar pattern of epitope focusing and spreading may occur during the transition from isolated monosymptomatic demyelinating syndromes (a group of distinct clinical disorders with variable rates of progression to MS) to clinically-defined MS (18). Thus, elucidation of the cellular and molecular mechanisms driving the epitope spreading process are critical to the design of efficient therapies for treating chronic inflammatory autoimmune diseases. However, until recently it has been unclear whether T cells specific for endogenous myelin epitopes play a significant pathologic role in tissue damage during the relapsing clinical episodes, or whether epitope spreading is simply an epi-phenomenon without pathologic consequence. We have therefore explored the immunopathological consequences of epitope spreading in R-EAE induced in SJL mice by both the highly encephalitogenic PLP139-151 epitope and the weakly encephalitogenic MBP84-104 epitope.

Multiple approaches were employed to demonstrate that the responses to the endogenous myelin epitopes were primary responsible for mediating the relapsing clinical episodes. First, we used serial transfer assays in which splenic T cells from SJL mice which had recovered from the acute clinical episode of R-EAE were re-activated *in vitro* with a panel of myelin peptides and then tested for their ability to mediate disease after adoptive transfer to naïve recipient mice. These studies indicated that splenic T cells from mice which had recovered from PLP139-151-induced R-EAE could transfer clinical disease following reactivation with either the initiating PLP139-151 peptide or with the relapse-associated PLP178-191 peptide (17). Similarly, splenocytes from mice with MBP84-104-induced R-EAE, activated *in vitro* with PLP139-151, can serially transfer R-EAE to naïve mice (19). These studies indicate that the T cells activated to endogenous myelin epitopes released during acute myelin destruction have encephalitogenic potential . Secondly, we show that T cells specific for the relapse-associated epitopes are demonstrable in the CNS of mice during disease remission (19). Thus, T cells recovered from the CNS of mice with PLP139-151-induced R-EAE during the primary relapse proliferated to PLP178-191 and T cells from mice in relapse from MBP84-104-induced R-EAE proliferated to PLP139-151. The strongest evidence for concluding that responses to spread epitopes are predominant during disease relapses comes from peptide-specific tolerance studies. Induction of peptide-specific tolerance to the relapse-associated epitope alone, either prior to disease induction or after the acute phase, blocks disease progression as assessed by a decreased relapse rate (19). An example of the effects of tolerance on PLP139-151-induced EAE is shown in Figure 2. As seen in Figure 2A, pre-tolerization (seven days prior to disease

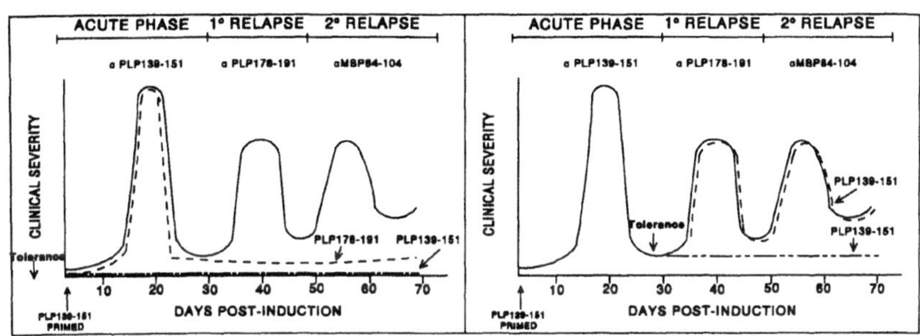

Figure 2. PLP178-191-specific tolerance prior to disease induction (Left Panel) or during remission (Right Panel) blocks disease progression in active PLP139-151-induced R-EAE.

induction) with the initiating PLP139-151 peptide results in total protection from disease. Pre-tolerization with the relapse-associate PLP178-191 epitope had no effect on acute disease, but inhibits development of the primary relapse suggesting that the response to the relapse-associated PLP178-191 epitope is principally responsible for the primary disease relapse. In confirmation, induction of tolerance following recovery from acute PLP139-151-induced EAE (Figure 2B) with the relapse-associated PLP178-191 epitope alone, but not with PLP139-151, significantly inhibited development of relapsing disease episodes. Collectively, these results indicate that clinically relevant epitope spreading takes place in a hierarchical order of peptide dominance and plays a critical role in disease progression.

FUNCTIONAL ROLE OF EPITOPE SPREADING IN MEDIATING CHRONIC TMEV-INDUCED DEMYELINATING DISEASE

TMEV-IDD is the most relevant of the available virus-induced animal models of immune-mediated demyelination (11) and serves as an excellent system in which to assess the potential contribution of anti-myelin autoimmune responses to initiation and progression of clinical demyelination. SJL mice develop a persistent CNS infection accompanied by a chronic, progressive, T cell-mediated inflammatory demyelinating disease presenting clinically initially as a problem with gait spasticity at about 30-35 days post-infection and progressing to total hind limb paralysis (20-22). Interestingly, TMEV persists in the CNS for the lifetime of the host by continuous, low-level virus replication predominantly in APCs, *i.e.* microglia and infiltrating macrophages (23).

There is considerable evidence supporting the concept that TMEV-IDD is a CD4$^+$ T cell-mediated immunopathology. Depletion of CD4$^+$, but not CD8$^+$, T cells prevents demyelination in SJL mice (24). Pathologic disease is equivalent in β2-microglobulin-deficient SJL mice lacking CD8$^+$ T cells and their +/- littermates (manuscript in preparation) indicating that CD4$^+$ T cells are sufficient to mediate disease. Disease susceptibility correlates with the development of TMEV-specific, MHC class II-restricted delayed-type hypersensitivity (DTH) responses (25-27), and is characterized by the predominant production of Th1-like cytokines (28-30) and IgG2a anti-virus antibody (31). In addition, transfer of TMEV-specific polyclonal T cells or long-term CD4$^+$ T cell lines leads to an increase in disease incidence and severity (24).

Time-course studies comparing the development of T cell responses to both virus and myelin epitopes have shown that responses to TMEV epitopes (VP270-86 and VP324-37) are clearly demonstrable by 5-7 days post-infection and persist for as long as 250 days (26). In contrast, autoreactivity to myelin epitopes is not detected prior to the onset of clinical disease which occurs between 30-35 days post-infection (32,33). Induction of peripheral tolerance to mouse spinal cord homogenate (MSCH - a heterogeneous mixture of multiple neuroantigens) which is effective in preventing clinical and histologic signs of R-EAE induced by a variety of myelin proteins and peptides does not affect the clinical onset of demyelination and the accompanying virus-specific T cell and antibody responses in TMEV-infected SJL/J mice (34). In contrast, tolerance to intact TMEV virions coupled to syngeneic splenocytes, which specifically anergizes virus-specific Th1 responses (28,29), results in a dramatic reduction in the incidence and severity of demyelinating lesions and clinical disease in SJL/J mice subsequently infected with TMEV (35). Collectively, these results strongly support the hypothesis that TMEV-IDD is *initiated* by virus-specific T cells targeting viral epitopes presented by persistently infected CNS-resident APCs. It is likely that pro-inflammatory cytokines and chemokines produced by TMEV-specific T cells (36,37) then result in the attraction and activation of microglia and mononuclear inflammatory cells which mediate the early phases of myelin destruction by bystander mechanisms.

Although T cell reactivity to myelin epitopes is not observed in the early stages of TMEV-IDD, a temporal analysis of T cell responses to a panel of encephalitogenic myelin epitopes has demonstrated that myelin-specific autoreactive T cells arise during the later stages of TMEV-IDD. Beginning 50-60 days post-infection (~ 3 weeks after the onset of clinical disease), T cell proliferative and DTH responses to the highly immunodominant PLP139-151 epitope develop. As disease progresses, DTH responses to multiple myelin epitopes arise (8,38). Approximately one month following the appearance of the PLP139-151-specific T cell response, responses to PLP56-70 and MOG92-106 are demonstrable. This is followed by the development of T cells reactive to PLP178-191 within five months post-infection, and responses to MBP84-104 arise another month later. Responses to the cryptic PLP104-117 epitope were not seen in TMEV-infected mice. Thus, chronic CNS inflammation initiated by virus-specific T cell responses leads to the induction and persistence of autoreactive T cells during the chronic stage of TMEV-IDD. A schematic of the pattern of epitope spreading in TMEV-IDD is depicted in Figure 3.

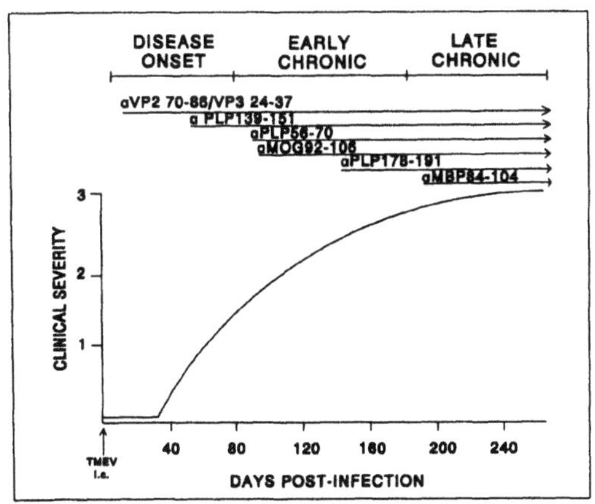

Figure 3. Schematic of development of virus and myelin-specific T cell responses in .TMEV-induced demyelinating disease.

Current evidence indicates that the myelin-specific autoimmune responses which arise during chronic TMEV-IDD are activated via epitope spreading rather than as a result of shared epitopes (molecular mimicry) between TMEV and the myelin proteins. The difference in the temporal appearance of T cell reactivity to TMEV vs. the involved myelin epitopes does not support a molecular mimicry hypothesis. In addition, there is no sequence homology between TMEV and the major myelin proteins as determined by computer searches (8). Also, there is no crossreactivity observed when panels of polyclonal lymph node T cells and T cell hybridoma clones specific for the immunodominant TMEV epitopes (VP1233-250, VP270-86, and VP324-37) are tested with a panel of encephalitogenic myelin proteins, or PLP epitopes (PLP56-70, PLP104-117, PLP139-151, and PLP178-191), MBP84-104, MOG92-106, or by a mouse spinal cord homogenate (MSCH). Similarly, representative PLP139-151-specific T cell hybrid clones were not activated in a cross-reactive fashion by intact TMEV, GST-fusion proteins encompassing the VP1, VP2 or VP3 capsid proteins, or by the immunodominant VP capsid epitopes.

Importantly, recent studies (manuscript in preparation) have indicated that the autoreactive T cells arising during chronic TMEV-IDD play the major pathologic role in chronic disease progression. The induction of peripheral tolerance to MP4 (a recombinant fusion protein comprising the 21.5 kD isoform of human MBP and a recombinant variant of human PLP) at 45 days post-infection (just prior to the initial appearance of PLP139-151-specific responses) ameliorates further disease progression. In addition, PLP139-151-specific IFN-γ-producing Th1 cells infiltrate the CNS of chronically infected mice. *Collectively, these results support a model wherein virus-specific T cells drive early myelin destruction, leading to the de novo activation and recruitment of myelin-specific autoreactive T cells which principally mediate chronic disease pathology.*

ENDOGENOUS PRESENTATION OF VIRUS AND MYELIN EPITOPES BY CNS-RESIDENT APCs IN TMEV-INFECTED MICE

We have also examined the ability of APCs isolated from the CNS target organ to endogenously present virus and myelin epitopes at various stages of R-EAE and TMEV-IDD (15,39). Our analyses have shown that macrophage/microglia isolated from the CNS of mice in remission from PLP139-151-induced R-EAE can endogenously present multiple myelin epitopes (PLP139-151, PLP178-191 and MBP84-104) to specific T cells lines. Thus, non-crossreactive epitopes on both the initiating PLP antigen and on MBP are acquired by the CNS-resident APCs during ongoing R-EAE. CNS APCs from TMEV-infected SJL mice have the ability to endogenously process and present several virus epitopes, including VP270-86 and VP324-37, at both acute and chronic stages of the disease. This supports the conclusion that CNS-infiltrating macrophages/resident microglia are persistently infected with virus and can readily present viral epitopes either directly and/or by phagocytosing debris containing virus antigens. Relevant to the initiation of myelin-specific autoimmune responses in chronic TMEV-IDD, only CNS APCs isolated from TMEV-infected mice with pre-existing myelin damage ($\geq$60-80 days post-infection), but not those isolated from naïve mice or mice at the early stages of disease, were able to endogenously present a variety of PLP epitopes, including PLP56-70, PLP104-117, PLP139-151 and PLP178-191, to specific T cell hybridomas and Th1 lines. This timing is roughly equivalent to the initial appearance of T cells specific for the immunodominant PLP139-151 epitope. However, this temporal analysis of the appearance of various myelin epitopes (38) indicated that these epitopes are available on CNS APCs well before peripheral T cell responses to most of these epitopes are demonstrable (see Figure 3). Phenotypic analysis of the F4/80[+] CNS-resident cells capable of presenting endogenous myelin epitopes indicates that these cells are composed of approximately 60% microglia (CD45^{intermediate}) and 30% infiltrating macrophages (CD45^{high}). These cells also express high levels of MHC class II (I-A^{s}) and B7-1 and B7-2 costimulatory molecules and are thus in all aspects professional APCs (15). Collectively this data provides a strong mechanistic basis for epitope spreading by showing that various encephalitogenic myelin epitopes can be processed and presented by APCs in the CNS target organ.

SUMMARY

These results support a model of epitope spreading (Figure 4) wherein localized virus-specific T cell-mediated inflammatory processes lead to the recruitment/activation of CNS-resident APCs which can serve both as effector cells for myelin destruction and as APCs which efficiently process/present endogenous self epitopes to autoreactive T cells. Thus, inflammatory responses induced by viruses which trigger pro-inflammatory Th1 responses *and* have the ability to persist in genetically susceptible hosts, may lead to chronic organ-specific autoimmune disease via epitope spreading. Regardless of the specificity of the T cells (myelin peptides in R-EAE or TMEV epitopes in TMEV-IDD) responsible for initiating myelin destruction, epitope spreading plays an important contributory role in the chronic disease process in genetically susceptible SJL mice. Epitope spreading has obvious important implications to the design of antigen-specific therapies for the potential treatment of MS and other autoimmune diseases. This process indicates that autoimmune diseases are evolving pathologies and that the specificity of the effector autoantigen-specific T cells varies during the chronic disease process.

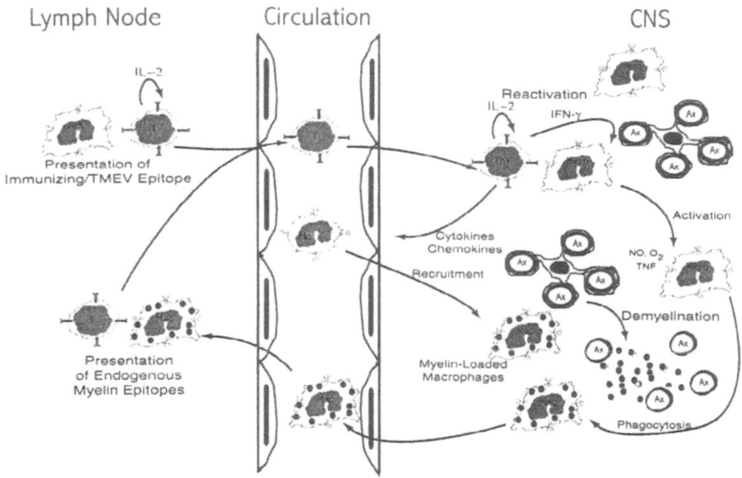

Figure 4. Model of epitope spreading in chronic CNS inflammatory demyelinating diseases.

REFERENCES

1. Wekerle H: Experimental autoimmune encephalomyelitis as a model of immune-mediated CNS disease. Curr Opin Neurobiol 3:779-784, 1993

2. Ota K, Matsui M, Milford EL, Mackin GA , Weiner IIL, Hafler DA: T-cell recognition of an immunodominant myelin basic protein epitope in multiple sclerosis. Nature 346:183-187, 1990

3. Bernard CC, de Rosbo NK: Immunopathological recognition of autoantigens in multiple sclerosis. Acta Neurologica 13:171-178, 1991

4. de Rosbo NK, Hoffman M, Mendel I, Yust I, Kaye J, Bakimer R, Flechter S, Abramsky O, Milo R, Karni A, Ben-Nun A: Predominance of the autoimmune response to myelin oligodendrocyte glycoprotein (MOG) in multiple sclerosis: reactivity to the extracellular domain of MOG is directed against three main regions. Eur J Immunol 27:3059-3069, 1997

5. Kurtzke JF: Epidemiologic evidence for multiple sclerosis as an infection. Clin Microbiol Rev 6:382-427, 1993

6. Scherer MT, Ignatowicz L, Winslow GM, Kappler JW, Marrack P: Superantigens: bacterial and viral proteins that manipulate the immune system. Ann Rev Cell Biol 9:101-128, 1993

7. Fujinami RS, Oldstone MB: Amino acid homology between the encephalitogenic site of myelin basic protein and virus: mechanism for autoimmunity. Science 230:1043-1045, 1985

8. Miller SD, Vanderlugt CL, Begolka WS, Pao W, Yauch RL, Neville KL, Katz-Levy Y, Carrizosa A, Kim BS: Persistent infection with Theiler's virus leads to CNS autoimmunity via epitope spreading. Nature Med 3:1133-1136, 1997

9. Horwitz MS, Bradley LM, Harbetson J, Krahl T, Lee J, Sarvetnick N: Diabetes induced by Coxsackie virus: initiation by bystander damage and not molecular mimicry. Nature Med 4:781-786, 1998

10. McRae BL, Kennedy MK, Tan LJ, Dal Canto MC, Miller SD: Induction of active and adoptive chronic-relapsing experimental autoimmune encephalomyelitis (EAE) using an encephalitogenic epitope of proteolipid protein. J Neuroimmunol 38:229-240, 1992

11. Miller SD, Gerety SJ: Immunologic aspects of Theiler's murine encephalomyelitis virus (TMEV)-induced demyelinating disease. Semin Virol 1:263-272, 1990

12. Miller SD, McRae BL, Vanderlugt CL, Nikcevich KM, Pope JG, Pope L, Karpus WJ: Evolution of the T cell repertoire during the course of experimental autoimmune encephalomyelitis. Immunol Rev 144:225-244, 1995

13. Vanderlugt CL, Miller SD: Epitope spreading. Curr Opin Immunol 8:831-836, 1996

14. Vanderlugt CL, Karandikar NJ, Bluestone JA, Miller SD: The functional significance of epitope spreading and its regulation by costimulatory interactions. Immunol Rev 164:63-72, 1998

15. Katz-Levy Y, Neville KL, Girvin AM, Vanderlugt CL, Pope JG, Tan LJ, Miller SD: Endogenous processing of self myelin epitopes by CNS-resident APCs in Theiler's virus-infected mice. J Clin Invest 104:599-610, 1999

16. Lehmann PV, Sercarz EE, Forsthuber T, Dayan CM, Gammon G: Determinant spreading and the dynamics of the autoimmune T-cell repertoire. Immunol Today 14:203-208, 1993

17. McRae BL, Vanderlugt CL, Dal Canto MC, Miller SD: Functional evidence for epitope spreading in the relapsing pathology of experimental autoimmune encephalomyelitis. J Exp Med 182:75-85, 1995

18. Tuohy VK, Yu M, Weinstock-Guttman B, Kinkel RP: Diversity and plasticity of self recognition during the development of multiple sclerosis. J Clin Invest 99:1682-1690, 1997

19. Vanderlugt CL, Eagar TN, Neville KL, Nikcevich KM, Bluestone JA, Miller SD: Pathologic role and temporal appearance of newly emerging autoepitopes in relapsing experimental autoimmune encephalomyelitis. J Immunol 164:670-678, 2000

20. Lipton HL: Theiler's virus infection in mice: An unusual biphasic disease process leading to demyelination. Infect Immun 11:1147-1155, 1975

21. Dal Canto MC, Lipton HL: Primary demyelination in Theiler's virus infection. An ultrastructural study. Lab Invest 33:626-637, 1975

22. Lehrich JR, Arnason BGW, Hochberg F: Demyelinative myelopathy in mice induced by the DA virus. J Neurol Sci 29:149-160, 1976

23. Lipton HL: Persistent Theiler's murine encephalomyelitis virus infection in mice depends on plaque size. J Gen Virol 46:169-177, 1980

24. Gerety SJ, Rundell MK, Dal Canto MC, Miller SD: Class II-restricted T cell responses in Theiler's murine encephalomyelitis virus (TMEV)-induced demyelinating disease. VI. Potentiation of demyelination with and characterization of an immunopathologic CD4[+] T cell line specific for an immunodominant VP2 epitope. J Immunol 152:919-929, 1994

25. Clatch RJ, Melvold RW, Miller SD, Lipton HL: Theiler's murine encephalomyelitis virus (TMEV)-induced demyelinating disease in mice is influenced by the H-2D region: correlation with TMEV-specific delayed-type hypersensitivity. J Immunol 135:1408-1414, 1985

26. Clatch RJ, Lipton HL, Miller SD: Characterization of Theiler's murine encephalomyelitis virus (TMEV)-specific delayed-type hypersensitivity responses in TMEV-induced demyelinating disease: correlation with clinical signs. J Immunol 136:920-927, 1986

27. Clatch RJ, Lipton HL, Miller SD: Class II-restricted T cell responses in Theiler's murine encephalomyelitis virus (TMEV)-induced demyelinating disease. II. Survey of host immune responses and central nervous system virus titers in inbred mouse strains. Microb Pathogen 3:327-337, 1987

28. Peterson JD, Karpus WJ, Clatch RJ, Miller SD: Split tolerance of Th1 and Th2 cells in tolerance to Theiler's murine encephalomyelitis virus. Eur J Immunol 23:46-55, 1993

29. Karpus WJ, Peterson JD, Miller SD: Anergy *in vivo*: Down-regulation of antigen-specific CD4$^+$ Th1 but not Th2 cytokine responses. Int Immunol 6:721-730, 1994

30. Begolka WS, Miller SD: Cytokines as intrinsic and exogenous regulators of pathogenesis in experimental autoimmune encephalomyelitis. Res Immunol 149:771-781, 1998

31. Peterson JD, Waltenbaugh C, Miller SD: IgG subclass responses to Theiler's murine encephalomyelitis virus infection and immunization suggest a dominant role for Th1 cells in susceptible mouse strains. Immunol 75:652-658, 1992

32. Barbano RL, Dal Canto MC: Serum and cells from Theiler's virus-infected mice fail to injure myelinating cultures or to produce in vivo transfer of disease. The pathogenesis of Theiler's virus-induced demyelination appears to differ from that of EAE. J Neurol Sci 66:283-293, 1984

33. Miller SD, Clatch RJ, Pevear DC, Trotter JL, Lipton HL: Class II-restricted T cell responses in Theiler's murine encephalomyelitis virus (TMEV)-induced demyelinating disease. I. Cross-specificity among TMEV substrains and related picornaviruses, but not myelin proteins. J Immunol 138:3776-3784, 1987

34. Miller SD, Gerety SJ, Kennedy MK, Peterson JD, Trotter JL, Tuohy VK, Waltenbaugh C, Dal Canto MC, Lipton HL: Class II-restricted T cell responses in Theiler's murine encephalomyelitis virus (TMEV)-induced demyelinating disease. III. Failure of neuroantigen-specific immune tolerance to affect the clinical course of demyelination. J Neuroimmunol 26:9-23, 1990

35. Karpus WJ, Pope JG, Peterson JD, Dal Canto MC, Miller SD: Inhibition of Theiler's virus-mediated demyelination by peripheral immune tolerance induction. J Immunol 155:947-957, 1995

36. Begolka WS, Vanderlugt CL, Rahbe SM, Miller SD: Differential expression of inflammatory cytokines parallels progression of central nervous system pathology in two clinically distinct models of multiple sclerosis. J Immunol 161:4437-4446, 1998

37. Hoffman LM, Fige BT, Begolka WS, Miller SD, Karpus WJ: Central nervous system chemokine expression during Theiler's virus-induced demyelinating disease. J Neurovirol 5:635-642, 1999

38. Katz-Levy Y, Neville KL, Padilla J, Rahbe SM, Begolka WS, Girvin AM, Olson JK, Vanderlugt CL, Miller SD: Temporal development of autoreactive Th1 responses and endogenous antigen presentation of self myelin epitopes by CNS-resident APCs in Theiler's virus-infected mice. J Exp MedSubmitted, 2000

39. Pope JG, Vanderlugt CL, Lipton HL, Rahbe SM, Miller SD: Characterization of and functional antigen presentation by central nervous system mononuclear cells from mice infected with Theiler's murine encephalomyelitis virus. J Virol 72:7762-7771, 1998

MULTIPLE SCLEROSIS AND GENE EXPRESSION PROFILING

Lawrence Steinman

Dept. of Neurological Sciences
Stanford University
Stanford, CA 94305

1. INTRODUCTION

In gene microarray studies on MS brain, my lab has been studying mRNA from active MS brain lesions from MS brains, and comparing them to three control brains, as well as four brains with Huntington's Disease, and four brains with Parkinson's Disease.[1,2] Using a 6800 gene Affymetrix microarray, large increases in transcription of genes encoding key components of the immune response have been seen.[1,2,3]

2. THE PRESUMED PATHOPHYSIOLOGY OF DEMYELINATING DISEASE AND WHAT GENE MICROARRAYS CAN TEACH US

Multiple sclerosis is an inflammatory demyelinating disease of CNS white matter. Evidence points to a misdirected immune response against myelin antigens. A widely accepted view of the process of demyelination reveals that T cells, immunoglobulin and complement play a role in pathogenesis. Initially, circulating autoreactive myelin-specific CD4+ Th1 cells are activated in the periphery by non-self antigens with a resemblance to CNS myelin proteins. These circulating T cells are presumably captured on vascular endothelium in the brain first by selectins, and then by ligand-receptor interactions between integrin molecules. T cells migrate through the endothelium in response to chemotactic signals, and metalloproteinases or other matrix degrading enzymes facilitate T cell penetration through the basement lamina. Microglial cells and astrocytes reactivate T cells locally in the CNS by presentation of myelin antigens complexed with MHC class II molecules. T cells stimulate macrophages by secretion of inflammatory cytokines, and macrophages in turn show increased phagocytic activity and release free oxygen radicals, nitric oxide metabolites, proteases, arachidonic acid derivatives and complement components that damage myelin. Autoantibodies against myelin antigens, such as myelin basic protein and myelin oligodendroglial glycoprotein also have an important role in demyelination. Many of these gene products involved in this scenario have been demonstrated at the sites of lesions by standard methods of immunohistochemistry, in-situ hybridization, and reverse-transcriptase based PCR. Some understanding of the sequence of events in the pathogenesis of the MS lesion has helped to direct the development of specific therapies.[4,5]

The rapid pace of human genome sequencing with the accompanying development of technologies to identify novel genes, has already revealed sequences for approximately one half of the estimated 100,000 genes expressed in the human genome. Understanding the integrated function of these genes is a difficult task. One approach relies on comparative studies of gene expression in healthy and diseased tissue, for evaluating processes *in vivo* for clues of functions gone awry in the disease state. The pathogenesis of MS is dependent on which of these 100,000 genes are transcribed and ultimately expressed, and the location, amplitude and timing of events. Microarray technologies allow the transcription levels of thousands of genes to be measured simultaneously, potentially on a whole genome scale.[6] Several different types of microarrays have been developed based on cDNA clones, PCR products, or oligonucleotides immobilized or synthesized on a solid support.

The Affymetrix GeneChip™ consists of a high-density oligonucleotide array synthesized directly onto glass slides by photochemical methods. Each "synthesis feature" or "probe cell" of a particular gene contains more than 10^7 copies of its 25-mer DNA oligonucleotide sequence. As many as 400,000 "synthesis features", each of a 24 x 24 micron size, can be placed on a 1.28 x 1.28 cm chip. Each gene on the array is represented by a series of 20-60 probes or features that span the sequence. The "synthesis features" are organized in pairs, consisting of a perfect match (PM) oligonucleotide, and a mismatch oligonucleotide (MM) immediately below. The MM oligonucleotide serves as a control for non-specific hybridization specificity and has a single base change at the middle position. Samples for hybridization are prepared by isolation of polyA+ mRNA which is made into cDNA, and then converted into biotinylated complimentary RNA (cRNA). The biotinylated cRNA is fragmented and hybridized to the array. After washing and staining, the array is scanned with a confocal laser microscope. The presence or absence of a particular RNA is determined from the differences and ratios of the PM and MM fluorescence intensities (see Methods). For a particular gene, the signal intensity is proportional to the amount of labeled cRNA bound to the DNA oligonucleotide probes on the chip. The dynamic range is wide and signal intensity is fairly linear over a 10^3-10^4-fold range of concentrations to allow the relative concentrations of different RNAs in a population to be estimated. The chips can detect message at an abundance of 1:100,000, which corresponds to approximately 3-5 copies per cell.[6] The use of short specific oligonucleotides allows discrimination between members of gene families. Only sequence information is required to produce the chip, and there is no handling of clones or PCR products involved.[7]

3. MS AND ITS ANIMAL COUNTERPART, EAE

I have recently reviewed the pros and cons of using the EAE model as a predictor of success for new drugs in MS[5] (Table 1). I wrote in conclusion: "Though EAE has many features that reflect what is known about the pathophysiology of MS, there are many discrepancies between the pathology of EAE and MS. Therefore, extrapolations must be made with caution when predicting what might happen in MS, based on results obtained in the EAE model. Results of ongoing clinical trials, targeting alpha 4 integrin with humanized antibodies, metalloproteases with small molecule inhibitors, and T cells reactive to myelin antigens with humanized antibodies and altered peptide ligands, should allow us to judge the utility of EAE as a model to predict therapeutic efficacy for MS. So far, the development of the approved MS drug, Copaxone, based on its efficacy in EAE, and the success of beta interferon in treating EAE, must be balanced with the failures of anti-TNF-α and gamma-IFN in MS, after their successful use in various models of EAE. It is too early to judge whether amelioration of EAE will be a reliable predictor of success in treating MS."

Table 1   Comparisons Between Multiple Sclerosis and EAE

	MS	EAE
<u>Clinical Presentation</u>		
relapses and remissions	present	present
paralysis	present	present
ataxia	present	present
visual impairment	present	present
Genetics		
MHC linked susceptibility	yes	yes
Females more susceptible	yes	yes
Pathology in Lesions		
T cells reactive to myelin	present	present
Antibodies to myelin	present	present
α4 integrin, complement	present	present
TNFα, γIFN,	present	present
Demyelination	present	present
Axonal dystrophy	present	present
<u>Therapy</u>		
βIFN, systemic	worsens	cures
anti-TNFα systemic	worsens	cures
IL-4 transduced T cells	not done	cures
TNFα Transduced T cells	not done	worsens
Copaxone	improves	cures
βIFN	improves	improves

Despite these caveats about EAE as a model system for MS we are studying transcriptional profiles of EAE brain during acute attacks, during relapse, and also compare transcriptional profiles in EAE brains after therapy with altered peptide ligands.[5]

4. REFERENCES

1. Lock C, Steinman L, Oksenberg J, Raine C, and Heller R.   Large-scale expression analysis of MS lesions using gene microarrays. *Neurology* **52**, A439, (1999).
2. Karpuj MV and Steinman L. Transcriptional profile of Huntington disease brain with gene microarrays. *Society for Neuroscience Abstracts* **25**, part I, 541, (1999).
3. Lock C, Oksenberg J and Steinman L.   The role of TNFa and lymphotoxin in demyelinating disease. *Ann Rhem Dis* **58**, I121-I128, (1999).
4. Steinman L   Multiple sclerosis: A coordinated immunological attack against myelin in the central nervous system. *Cell* **85**, 299-302, (1996)
5. Steinman L.   Assessment of the utility of animal models for multiple sclerosis and demyelinating disease in the design of rational therapy. *Neuron*, **24**, 511-514, (1999).

6. Heller, R. A. *et al.* Discovery and analysis of inflammatory disease-related genes using cDNA microarrays. *Proc Natl Acad Sci USA* **94**, 2150-2155 (1997).
7. Steinman L. Multiple mechanisms of multiple sclerosis. *Nature Med* **6**,15-16, (2000).

TREATMENT OF AUTOIMMUNITY BY INHIBITION OF T CELL COSTIMULATION

David I. Daikh and David Wofsy

Department of Medicine
University of California, San Francisco
Department of Veterans Affairs Medical Center
San Francisco, CA 94121

INTRODUCTION

Recent advances in our understanding of the mechanisms underlying immune responses have led to the development of new strategies designed to inhibit pathologic immune responses, such as autoimmunity, without severely compromising protective immune responses or causing serious toxic side effects. Many of these strategies are based on the two-signal model for T cell activation (1,2). The first signal occurs when the T cell receptor (TCR) recognizes an antigenic peptide displayed on the surface of an antigen-presenting cells (APC). The second signal is provided by other receptor-ligand pairs on T cells and APC. The presence or absence of this signal, also referred to as T cell costimulation, plays a critical role in determining whether antigen recognition through the TCR results in T cell activation or T cell unresponsiveness (1). Thus, the two-signal model implies that selective blockade of the second signal might render autoreactive T cells unresponsive in people with autoimmune diseases. As described below, this strategy has already been tested and has shown promise in animal models for autoimmunity, and it currently the subject of clinical investigation in humans.

INHIBITION OF THE INTERACTION BETWEEN B7 AND CD28

Upon interaction with antigen, APC are stimulated to express certain surface molecules that are not present on resting APC. Among these, the B7 molecules (B7-1 and B7-2) play a key role in providing the costimulatory second signal to T cells through their interaction with the T cell surface molecule CD28 (1,2). Consistent with the two-signal model for T cell activation, selective inhibition of the B7/CD28 interaction can induce antigen-specific T-cell unresponsiveness *in vitro* (2,3). Attempts to extend this observation to *in vivo* systems have taken advantage of the homology between CD28 and another T-cell surface molecule, designated CTLA-4. CTLA-4 appears to be a negative regulator of T cell function (4,5). It is expressed on activated T cells, and it binds B7 with considerably higher avidity than does CD28 (6,7). Therefore, a fusion protein consisting of the extracellular domain of CTLA-4 bound to an immunoglobulin $C\gamma1$ chain (CTLA4Ig) binds B7-1 and B7-2, blocks the interaction between the B7 molecules and CD28, and inhibits T-cell activation (3,8).

Because B7-1 and B7-2 are preferentially expressed on activated B cells and APC, it has been postulated that blockade of these molecules might preferentially affect T cells that are in the process of antigen recognition (1). In the setting of autoimmune disease, such

Mechanisms of Lymphocyte Activation and Immune Regulation VIII
Edited by Sudhir Gupta, Kluwer Academic/Plenum Publishers, 2001

blockade might provide a means of selectively inhibiting autoreactive T cells without affecting the majority of T cells. This hypothesis was tested by Milich *et al.* (9), who used CTLA4Ig to induce long-term unresponsiveness to an autoantigen in a transgenic model for autoantibody production. In similar studies, we examined the effects of CTLA4Ig in the NZB/NZW F_1 (B/W) murine model for systemic lupus erythematosus (SLE). Initially, we treated female B/W mice with murine CTLA4Ig for four months beginning at the onset of disease (10). Treatment prevented autoantibody production, retarded lupus nephritis, and significantly prolonged life (Figure 1). Subsequently, we showed that, even when treatment was delayed until severe disease had been established, CTLA4Ig had substantial beneficial effects (10). Although the benefits of treatment persisted for several months after discontinuation of CTLA4Ig, long-lasting self-tolerance was not established. Rather, lupus nephritis eventually recurred unless CTLA4Ig treatment was continued (11).

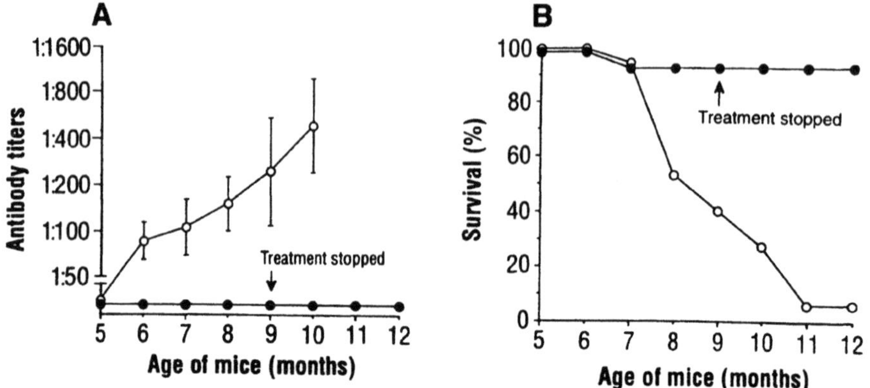

Figure 1: Female B/W mice were treated with CTLA4Ig (●) or a control mAb (○) from age 5 months to age 9 months (50 μg ip three times per week). (A) Geometric mean titer of antibodies to double-stranded DNA. (B) Percent survival. (Adapted with permission from reference 10: B.K. Finck, P.S. Linsley, and D. Wofsy, Treatment of murine lupus with CTLA4Ig, *Science* 265:1225-1227. Copyright 1994, American Association for the Advancement of Science.)

INHIBITION OF THE INTERACTION BETWEEN CD40 AND GP39

In addition to B7 and CD28, other receptor-ligand pairs can contribute to T cell costimulation. In particular, the interaction between CD40 on B cells and gp39 on T cells (also designated CD40-ligand) plays a key role in both T cell costimulation and B cell differentiation (12).

The B7/CD28 and CD40/gp39 pathways of costimulation are interrelated. For example, CD40 signaling enhances B7-dependent costimulation by increasing B7 expression on APC (12). Similarly, CD28-signaling can induce gp39 expression by T cells (13). To determine whether selective inhibition of CD40/gp39 interactions could, like CTLA4Ig, inhibit murine lupus, Mohan *et al.* (14) treated lupus-prone SWR/NZB (SNF$_1$) mice with three injections of mAb to gp39 at age 3 months, prior to the onset of autoimmune disease. Although treatment was brief, anti-gp39 delayed the development of lupus by several months without any further immunosuppressive therapy (Figure 2). However, the beneficial effect was not permanent, and the mice eventually went on to develop lupus nephritis. Early *et al.* (15) subsequently showed that chronic administration of anti-gp39 could retard murine lupus in B/W mice but, in this study, some of the mice eventually developed lupus nephritis despite ongoing therapy, at least in part because they developed an immune response to the hamster anti-gp39 mAb.

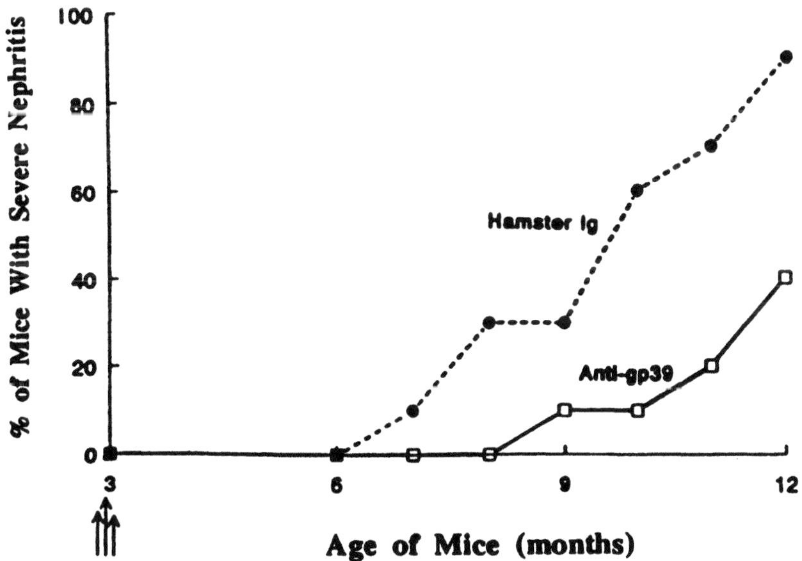

Figure 2: Incidence of severe nephritis (proteinuria >300 mg/dl) in female SNF_1 mice that received only three injections (250 μg/injection) of a hamster mAb to gp39 (□) or purified hamster Ig (●) at age three months. (Reprinted with permission from reference 14: C. Mohan, Y. Shi, J.D. Laman, and S.K. Datta, Interaction between CD40 and its ligand gp39 in the development of murine lupus nephritis, *J. Immunol.* **154**, 1470-1480. Copyright 1995, The American Association of Immunologists.)

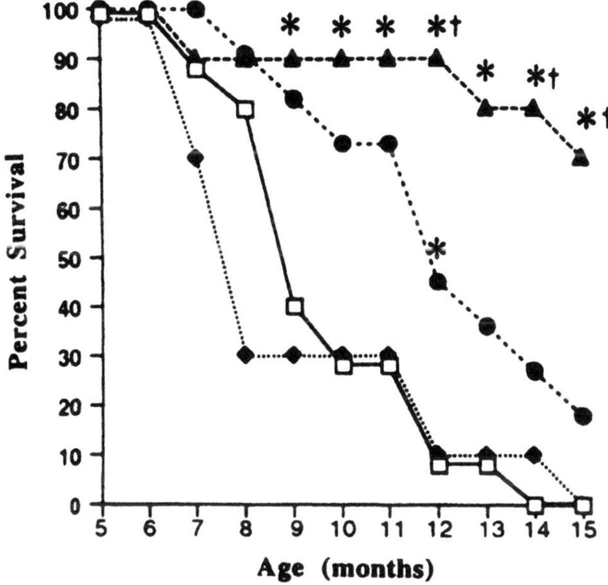

Figure 3: Female B/W mice were treated for 2 weeks at age 5 months with either CTLA4Ig alone (♦), mAb to gp39 alone (●), a combination of both CTLA4Ig and mAb to gp39 (▲), or control Ig (□). The asterisks (*) denote points at which there is a statistically significant difference from control mice (p<0.05). The daggers (†) denote points at which there is a statistically significant difference from mice that received mAb to gp39 alone (p<0.05). (Reprinted with permission from reference 11: D.I. Daikh, B.K. Finck, P.S. Linsley, D. Hollenbaugh, and D. Wofsy, Long-term inhibition of murine lupus by brief simultaneous blockade of the B7/CD28 and CD40/gp39 costimulation pathways, *J. Immunol.* **159**, 3104-3108. Copyright 1997, The American Association of Immunologists.)

SIMULTANEOUS BLOCKADE OF THE B7/CD28 AND CD40/GP39 PATHWAYS

The concept of blocking T cell costimulation in the setting of autoimmunity is based on the hope that this strategy might selectively induce unresponsiveness among T cells that are in the process of antigen recognition. Therefore, the eventual development of lupus nephritis in B/W mice treated with either CTLA4Ig or anti-gp39 was disappointing. However, it is possible that blockade of either pathway alone (B7/CD28 or CD40/gp39) may be insufficient to completely block T cell costimulation. Rather, simultaneous blockade of both pathways may be necessary to block costimulation. This hypothesis is supported by studies in an animal model for autoimmune oophoritis (16). In this model. neither CTLA4Ig nor anti-gp39 alone prevented proliferation of autoantigen-specific T cells but, when CTLA4Ig and anti-gp39 were combined, there was a marked synergistic effect that completely blocked expansion of autoreactive T cells.

To determine whether a similar synergistic effect could be achieved in murine lupus, we treated 5-month-old B/W females for 2 weeks with either CTLA4Ig alone, anti-gp39 alone, or both CTLA4Ig and anti-gp39 (11). The beneficial effects of the short course of either CTLA4Ig alone or anti-gp39 alone were relatively brief. However, when CTLA4Ig and anti-gp39 were combined, there was a marked synergistic effect (Figure 3). Ten months after cessation of therapy, 70% of the mice that received CTLA4Ig plus anti-gp39 were still alive, compared to only 18% survival among mice that received anti-gp39, and 0% survival among mice that received CTLA4Ig. These findings provide additional support for the prospect that brief blockade of T cell costimulation may provide long-lasting benefit in autoimmune diseases.

FUTURE DIRECTIONS

Based on the promising results in animal models for autoimmune disease, clinical trials have been initiated in humans in an effort to determine whether blockade of T cell costimulation will be effective in people with autoimmune diseases. In the first phase I trial of CTLA4Ig in humans, a brief series of four injections appeared to have a substantial beneficial effect in patients with psoriasis, a T-cell mediated autoimmune skin disease (17). A phase I trial of anti-gp39 has also been completed in humans (18). This study consisted of a single-dose dose-escalating examination of the pharmacokinetics and safety of anti-gp39 in patients with SLE, but it was not designed to examine efficacy. Phase II trials of CTLA4Ig and anti-gp39 are currently in progress.

ACKNOWLEDGMENTS

This work was supported by The Rosalind Russell Medical Research Center for Arthritis at UCSF and by the Department of Veterans Affairs.

REFERENCES

1. P. Marrack and J.W. Kappler, How the immune system recognizes the body, *Sci. Am.* **269**, 80-89 (1993).
2. F.A. Harding, J.G. McArthur, J.A. Gross, D.H. Raulet, and J.P. Allison, CD28-mediated signaling co-stimulates murine T cells and prevents induction of anergy in T cell clones, *Nature* **356**, 607-610 (1992).
3. P. Tan, C. Anasetti, J.A. Hansen, J. Melrose, M. Brunvand, J. Bradshaw, J.A. Ledbetter, and P.S. Linsley, Induction of alloantigen-specific hyporesponsiveness in human T lymphocytes by blocking interaction of CD28 with its natural ligand B7/BB1, *J. Exp. Med.* **177**, 165-173 (1993).
4. M. F. Krummel and J.P. Allison, CD28 and CTLA-4 have opposing effects on the response of T cells to stimulation, *J. Exp. Med.* **182**, 459-465 (1995).
5. P.J. Waterhouse, M. Penninger, E. Timms, A. Wakeham, A. Shahinian, K.P. Lee, C.B. Thompson, H. Griesser, and T. Mak, Lymphoproliferative disorders with early lethality in mice deficient in CTLA-4, *Science* **270**, 985-988 (1995).

6. J. A. Bluestone, New perspectives of CD28-B7-mediated T cell costimulation, *Immunity* **2**, 555-559 (1995).

7. P.S. Linsley, W. Brady, M. Urnes, L. Gorsmaire, N.K. Damle, and J.A. Ledbetter, CTLA-4 is a second receptor for the B cell activation antigen B7, *J. Exp. Med.* **174**, 561-569 (1991).

8. P.S. Linsley, P.M. Wallace, J. Johnson, M.G. Gibson, J.L. Greene, J.A. Ledbetter, C. Singh, and M.A. Tepper, Immunosuppression *in vivo* by a soluble form of the CTLA-4 T cell activation molecule, *Science* **257**, 792-795 (1992).

9. D. R. Milich, P.S. Linsley, J.L. Hughes, and J.E. Jones, Soluble CTLA-4 can suppress autoantibody production and elicit long term unresponsiveness in a novel transgenic model, *J. Immunol.* **153**, 429-435 (1994).

10. B.K. Finck, P.S. Linsley, and D. Wofsy, Treatment of murine lupus with CTLA4Ig, *Science* **265**, 1225-1227 (1994).

11. D. I. Daikh, B.K.Finck, P.S. Linsley, D. Hollenbaugh, and D. Wofsy, Long-term inhibition of murine lupus by brief simultaneous blockade of the B7/CD28 and CD40/gp39 costimulation pathways, *J. Immunol.* **159**, 3104-3108 (1997).

12. F. H. Durie, T.M. Foy, S.R. Masters, J.D. Laman, and R.J. Noelle, The role of CD40 in the regulation of humoral and cell-mediated immunity, *Immunol. Today* **15**, 406-411 (1994).

13. S.J. Klaus, I. Berberich, and E.A. Clark, CD40 and its ligand in the regulation of humoral immunity, *Semin. Immunol.* **6**, 279-286 (1994).

14. C. Mohan, Y. Shi, J.D. Laman, and S.K. Datta, Interaction between CD40 and its ligand gp39 in the development of murine lupus nephritis, *J. Immunol.* **154**, 1470-1480 (1995).

15. G.S. Early, W. Zhao, and C.M. Burns, Anti-CD40 ligand antibody treatment prevents the development of lupus-like nephritis in a subset of New Zealand Black x New Zealand White mice, *J. Immunol.* **157**, 3159-3164 (1996).

16. N.D. Griggs, S.S. Agersborg, R.J. Noelle, J.A. Ledbetter, P.S. Linsley, and K.S.K. Tung, The relative contribution of the CD28 and gp39 costimulatory pathways in the clonal expression and pathogenic acquisition of self reactive T cells, *J. Exp. Med.* **183**, 801-810 (1996).

17. J.R. Abrams, M.G. Lebwohl, C.A. Guzzo, B.V. Jegasothy, M.T. Goldfarb, B.S. Goffe, A. Menter, N.J. Lowe, G. Krueger, M.J. Brown, R.S. Weiner, M.J. Birkhofer, G.L. Warner, K.K. Berry, P.S. Linsley, J.G. Krueger, H.D. Ochs, S.L. Kelley, and S. Kang, CTLA4Ig-mediated blockade of T-cell costimulation in patients with psoriasis vulgaris, *J. Clin. Invest.* **103**, 1243-1252 (1999).

18. J.C. Davis, M.C. Totoritis, T.A. Sklenar, and D. Wofsy, Results of a phase I, single-dose, dose-escalating trial of a humanized anti-CD40L monoclonal antibody (IDEC-131) in patients with systemic lupus erythematosus (SLE), *Arthritis Rheum.* **42**, S281 (1999).

CYTOKINE BLOCKADE IN RHEUMATOID ARTHRITIS

Marc Feldmann, Ravinder N. Maini, Jan Bondeson, Peter Taylor
Brian M.J. Foxwell and Fionula M. Brennan

Kennedy Institute of Rheumatology
1 Aspenlea Road
Hammersmith
London W6 8LH U.K.

1.1. ABSTRACT

As we enter the 2000's it is clear that cytokine blockade is an effective therapeutic strategy for rheumatoid arthritis. In this brief review, we will review the rationale for anti TNFα therapy, the current status of therapy and focus on the regulation of TNFα production in rheumatoid synovium. New approaches to studying TNF regulation in RA and of elucidating the controversial role of T cells in this complex disease will be described.

2. THE RATIONALE FOR ANTI TNFα THERAPY IN RHEUMATOID ARTHRITIS

Over the past 15 years, many groups of workers, including Arend, Dayer, Duff, Koch, Lipsky as well as ourselves[1-7] have explored the cytokine expression in rheumatoid arthritis, in synovial fluid and synovium. These have documented a plethora of expressed cytokines.

At face value this data is confusing, and is open to many interpretations. Some workers lost interest and concluded that cytokines were poor targets for therapy on the assumption that these cytokines were independently regulated, and that blocking several of them was unrealistic.

Our approach was to delve deeper into the regulation of cytokines in rheumatoid synovium, and ascertain the mechanism of chronic cytokine upregulation in rheumatoid synovium. The technique was to develop a new way of culturing rheumatoid synovium. Prior to our work, it was common to culture rheumatoid synovium long term, passage the fibroblast like cells, and discard the haemopoietic cells which eventually disappeared, and then study these 'purified' cells. We believed that this approach was unrepresentative of the synovium *in vivo*, which is mostly comprised of leucocytes, chiefly macrophages, T lymphocytes, dendritic cells.

So we developed a simple *ex vivo* approach, to digest synovium, and plate out the dissociated synovial cells, *in vitro*. In the absence of extrinsic stimulation they produced an abundance of cytokines, with a prolonged expression pattern, unlike that found with antigen or mitogenic stimulation. This indicated to us that chronic cytokine production may be important in the disease, and also provided us with an *in vitro*[8] model to study the cytokine dysregulation occurring in RA synovium. As immunologists the obvious approach was to interrupt intercellular signalling with antibodies. In the light of evidence from J. Saklatvala and others that IL-1 was of major importance in initiating cartilage and bone damage, we studied IL-1 regulation. Brennan et al[9] reported that blocking TNFα dramatically downregulated IL-1 production (Figure 1). This was the first clue that TNFα had a special place in the cytokine 'network', and hence may be a therapeutic target.[10, 11] Subsequent studies revealed that anti TNFα downregulated GM-CSF, IL-6 and IL-8 expression in rheumatoid synovial cultures (reviewed[7]).

These *in vitro* studies highlighting the 'pivotal' role of TNFα in cytokine regulation were followed by other studies, immunohistological verification of TNFα and TNF receptor expression *in vivo*[12, 13] and then studies of the effects of anti TNFα in animal models of rheumatoid arthritis.[14, 15] These studies provided the rationale for the first clinical trials of anti TNFα antibody, initiated in May 1992. The efficacy of the chimaeric antibody used produced by Centocor, Inc, then termed cA2, now termed Infliximab or Remicade™, is shown in Figure 2.[11] Subsequent double blind randomized placebo controlled studies were performed and verified the importance of TNFα blockade as a therapeutic modality, and the effectiveness of this antibody, which is now licensed for clinical use in RA as a disease modifying agent.[16-21]

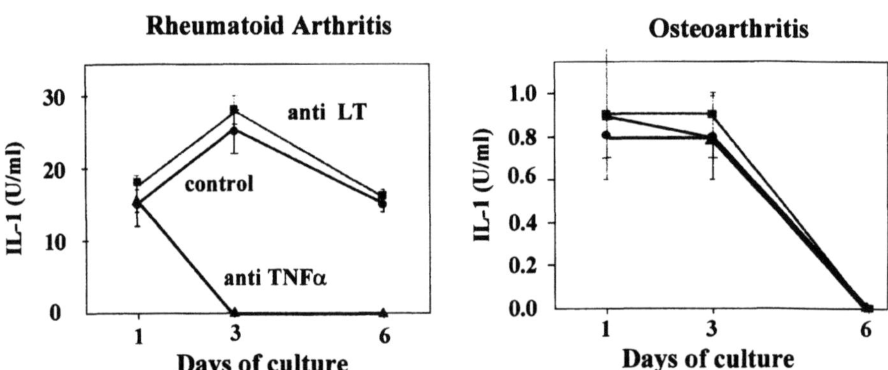

Figure 1. Inhibitory effect of anti-TNFα antibodies on synovial cell interleukin-1 production

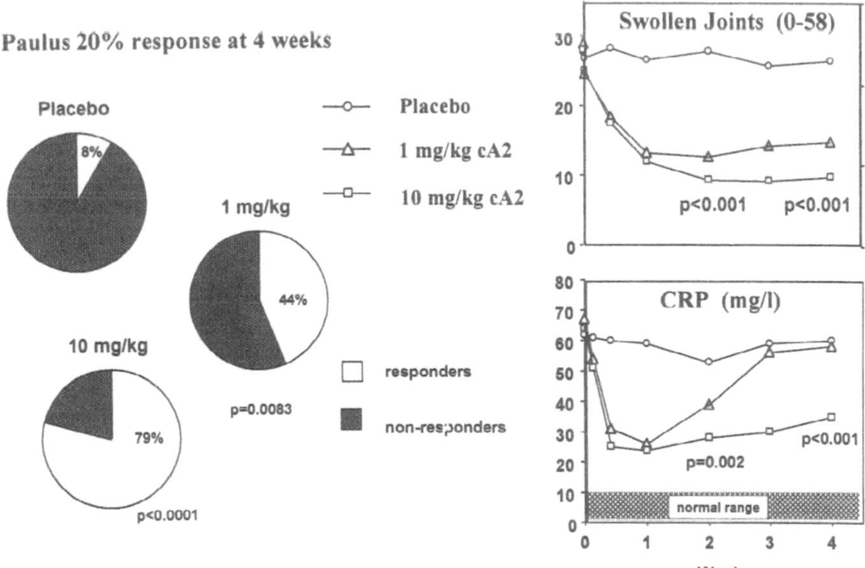

Figure 2. Responses to treatment with cA2.

3. WHAT DRIVES THE PRODUCTION OF TNFα IN RHEUMATOID SYNOVIUM?

With the accumulating evidence that excess TNFα was of major importance in the rheumatoid process, it was of interest to investigate the mechanism of TNFα upregulation in the synovium. We have approached this in several different ways.

One approach has been to investigate the intracellular signalling pathways which control rheumatoid TNFα production. How? We have used a number of approaches, one of the most interesting being adenoviruses expressing inhibitors of signalling pathways.

4. ADENOVIRUS IκBα INHIBITS TNFα AND MANY OTHER PROCESSES IN RHEUMATOID SYNOVIUM

While adenoviruses readily infect epithelial cells, their natural hosts, they also have the capacity to infect many other cells.[22, 23] Varying culture conditions, we have found that it is possible to infect macrophages at a high enough efficiency (>95%) to be able to block pathways.[24] With this approach we are able to demonstrate that while LPS induced TNFα was 80% inhibited by AdvIκBα, zymosan induced TNFα was not inhibited (Figure 3). [25] Other inducers of TNFα fell into these two groups, with UV light and PMA induced TNFα also NFκB dependent, but anti CD45 was independent of NFκB. [25, 26]

So, an interesting question was whether rheumatoid TNFα production is NFκB dependent. With AdvIκBα, synovial TNFα production was reduced by 70%. This suggests that the majority, but not all, of RA TNFα is NFκB dependent (Figure 4) [24] This prompted us to investigate how good a therapeutic target NFκB might be in rheumatoid tissue, by

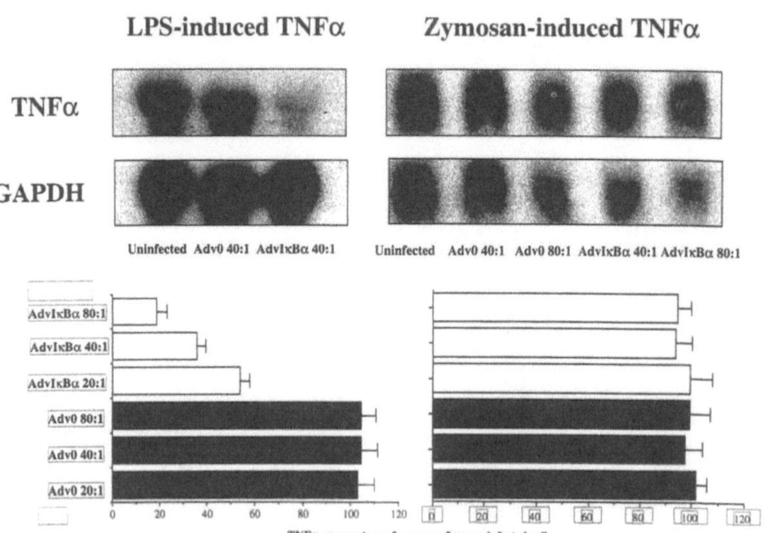

Figure 3. Effect of AdvIκBα on TNFα inhibition.

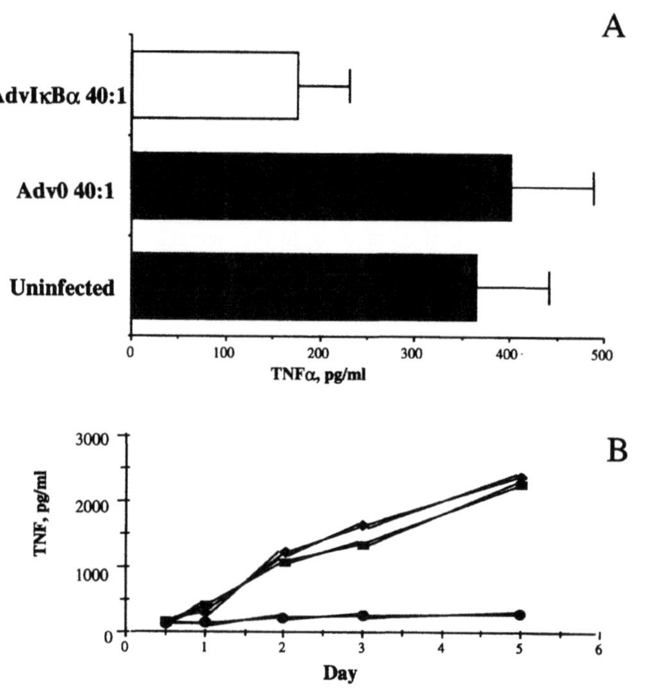

Figure 4. Effect of NFκB on RA TNFα production.

exploring what other changes may happen with AdvIκBα infection of rheumatoid synovial cultures. Marked inhibition (80%) of IL-6, partial inhibition of L-1β (60%) and IL-8 (50%) was noted indicating a widespread anti inflammatory effect compared to control virus with no insert, Ado. The effect on anti inflammatory cytokines was quite different, there was no inhibition in the rheumatoid cultures of the spontaneously produced anti inflammatory mediators including IL-10, IL-lRa, IL-l l and soluble TNF-R by AdvIκBα. Hence the effect of NFκB blockade on the overall equilibrium of pro-inflammatory to anti-inflammatory mediators, an aspect we have described to be a key component in the pathogenesis of RA.[7]

Metalloproteases are widely accepted to be of major importance in connective tissue breakdown, although with the increasing numbers of these described it is not known which ones are the most relevant.[27] We have studied the effect if AdvIκBα on MMP-1 (collagenase 1) MMP-3 (stromelysin) in rheumatoid synovium and MMP-13 in chondrosarcomas.

In all these instances, AdvIκBα reduced MMP production by 70-90%. In contrast there was no inhibition of TIMP-1, an inducible inhibitor of the MMP family.[28] Hence the overall effect of NFκB blockade on rheumatoid tissue revealed a potentially very beneficial therapeutic profile, and hence synovial NFκB appears to be a good therapeutic target.

Regrettably there are as yet no powerful and specific NFκB inhibiting drugs known, although there are many candidates proposed from existing drugs (e.g., aspirin). Many candidates being sought by high throughput screening using components of the NFκB activating pathways recently unravelled by the groups of Karin, Goeddel, Wallach etc.[29-31] As it is not known whether systemic blockade of NFκB would be safe, it appears that there is a case for trials of local NFκB blockade perhaps using adenoviruses expressing IκB. This proposal is supported by the work of Makorov et al[32] who have already shown in a rat model of arthritis that local infection of an adenovirus expressing a mutant, non degradable IκBα, induces improvements in the joint disease.

5. WHAT IS THE ROLE OF T LYMPHOCYTES IN RHEUMATOID ARTHRITIS?

This is a controversial question, with diametrically opposing views exposed in the literature, with even a book devoted to this topic.[33-35] Controversies exist because of lack of data, and we have been interested in this question, intrigued by the obvious presence of large numbers of T cells in RA synovium, the clear out HLA-DR genetic predisposition in caucasians with RA, the shared epitope, which suggests a role of T cells and antigen somewhere along the disease pathogenesis, and also by the relative lack of efficacy of therapy targetted at T cells and TCR signalling pathways in RA, compared to their efficacy in transplantation.[36]

As the excess TNFα production in RA is undoubtedly pathogenic, we investigated whether RA rheumatoid synovial TNF production was T cell dependent. This was attempted in 2 ways. By depleting T cells from rheumatoid synovial cell mixtures, it was found that macrophage TNFα production was markedly diminished. This suggested that T cells might be stimulating macrophages to produce proinflammatory cytokines, as had been proposed by Dayer and colleagues, based on their co-culture of T cells and macrophages.[37] We adopted their system, and used rheumatoid synovial T cells, fixed with monocytes purified from blood. It was found that rheumatoid T cells, freshly taken from synovium and not obviously stimulated, induced monocytes to produce TNFα. This stimulation was contact dependent, as shown by glutaraldehyde function as well as culture with cell impermeable culture inserts. By these two criteria RA synovial T cells did not appear to be 'innocent bystanders'.[38] We have followed up this work to study the pathways involved in rheumatoid T cell macrophage interactions. The results suggest that T cells in rheumatoid joints resemble the cytokine activated T cells first described in Unutmaz et al in 1994.[39]

Furthermore, the presence of cytokine activated T cells may explain the relative imbalance of TNFα and IL-10, with insufficient IL-10 produced in synovium to block TNFα. This is because T_{ck} can induce TNFα from macrophages, but not IL-10.[40] These T_{ck} may be important in perpetuating disease activity as they are dependent on a cocktail of cytokines (IL-6, TNFα and IL-15) which can all be made by macrophages, hence leading to a 'vicious cycle' (Figure 5).

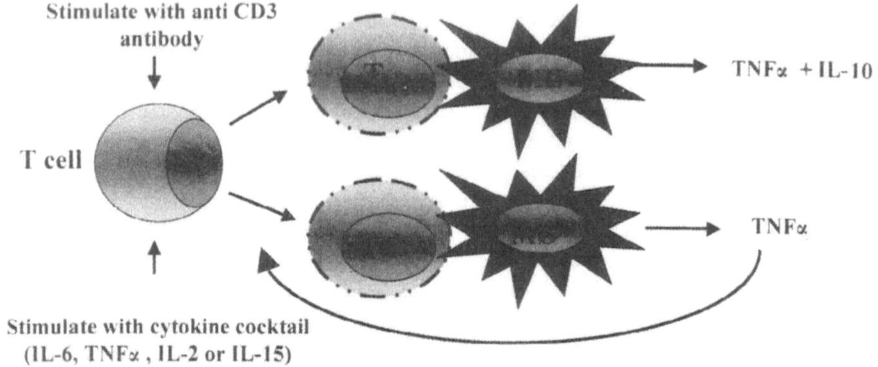

Figure 5. Cytokine disequilibrium induced by Tck.

6. CONCLUSIONS

New approaches to understanding the regulating TNFα production are now available. They offer the opportunity to work out how to mimic the benefit of anti TNFα biologicals without their expense.

7. REFERENCES

1. Arend, W. P. and Dayer, J. M. Cytokines and cytokine inhibitors or antagonists in rheumatoid arthritis. *Arthritis Rheum* 33, 305-315 (1990).
2. Roux-Lombard, P., Modoux, C., Vischer, T., Grassi, J. and Dayer, J. M. Inhibitors of interleukin 1 activity in synovial fluids and in cultured synovial fluid mononuclear cells. *J Rheumatol* 19, 517-523 (1992).
3. Koch, A. E. *et al.* Synovial tissue macrophage as a source of the chemotactic cytokine IL-8. *J Immunol* 147, 2187-2195 (1991).
4. Cush, J. J. *et al.* Elevated interleukin-10 levels in patients with rheumatoid arthritis. *Arthritis Rheum* 38, 96-104 (1995).
5. Di Giovine, F. S., Nuki, G. and Duff, G. W. Tumour necrosis factor in synovial exudates. *Ann Rheum Dis* 47, 768-772 (1988).

6. Hopkins, S. J., Humphreys, M. and Jayson, M. I. Cytokines in synovial fluid. I. The presence of biologically active and immunoreactive IL-1. *Clin Exp Immunol* 72, 422-427 (1988).

7. Feldmann, M., Brennan, F. M. and Maini, R. N. Role of cytokines in rheumatoid arthritis. *Annu Rev. Immunol.* 14, 397-440 (1996).

8. Buchan, G., Barrett, K., Turner, M., Chantry, D., Maini, R. N. and Feldmann, M. Interleukin-1 and tumour necrosis factor mRNA expression in rheumatoid arthritis: prolonged production of IL-1α. *Clin Exp Immunol* 73, 449-455 (1988).

9. Brennan, F. M., Chantry, D., Jackson, A., Maini, R. and Feldmann, M. Inhibitory effect of TNF alpha antibodies on synovial cell interleukin-1 production in rheumatoid arthritis. *Lancet* 2, 244-7 (1989).

10. Brennan, F. M., Maini, R. N. and Feldmann, M. TNFα—a pivotal role in rheumatoid arthritis? *Br J Rheumatol* 31, 293-298 (1992).

11. Feldmann, M. *et al.* Cytokine production in the rheumatoid joint: implications for treatment. *Ann Rheum Dis* 49, 480-486 (1990).

12. Chu, C. Q., Field, M., Feldmann, M. and Maini, R. N. Localization of tumor necrosis factor α in synovial tissues and at the cartilage-pannus junction in patients with rheumatoid arthritis. *Arthritis Rheum* 34, 1125-1132 (1991).

13. Deleuran, B. W., Chu, C.Q., Field, M., Brennan, F.M., Mitchell, T., Feldmann, M. and Maini, R.N.. Localization of tumor necrosis factor receptors in the synovial tissue and cartilage-pannus junction in patients with rheumatoid arthritis. Implications for local actions of tumor necrosis factor alpha. *Arthritis Rheum* 35, 1170-1178 (1992).

14. Williams, R. O., Feldmann, M. and Maini, R. N. Anti-tumor necrosis factor ameliorates joint disease in murine collagen-induced arthritis. *Proc Natl Acad Sci USA* 89, 9784-9788 (1992).

15. Thorbecke, G. J., Shah, R., Leu, C. H., Kuruvilla, A. P., Hardison, A. M. and Palladino, M. A. Involvement of endogenous tumour necrosis factor a and transforming growth factor β during induction of collagen type II arthritis in mice. *Proc Natl Acad Sci USA* 89, 7375-7379 (1992).

16. Elliott, M. J., Maini, R.N., Feldmann, M., Long-Fox, A., Charles, P., Katsikis, P., Brennan, R.M., Walker, J., Bijl, H., Ghrayeb, J. and Woody, J. Treatment of rheumatoid arthritis with chimeric monoclonal antibodies to tumor necrosis factor alpha. *Arthritis Rheum* 36, 1681-90 (1993).

17. Elliott, M. J., Maini, R.N., Feldmann, M., Kalden, J.R., Antoni, C., Smolen, J.S., Leeb, B., Breedveld, F.C., Macfarlane, J.D., Bijl, H. and Woody, J.N. Randomised double-blind comparison of chimeric monoclonal antibody to tumour necrosis factor alpha (cA2) versus placebo in rheumatoid arthritis. *Lancet* 344, 1105-10 (1994).

18. Maini, R.N., Breedveld, F.C., Kalden, J.R., Smolen, J.S., Davis, D., Macfarlane, J.D., Antoni, C., Leeb, B., Elliott, M.J., Woody, J.N., Schaible, T.F. and Feldmann, M. Randomized placebo-controlled trial of multiple intravenous infusions of anti-TNFα monoclonal antibody with or without weekly methotrexate in rheumatoid arthritis. *Arthritis Rheum.* 41, 1552-1563 (1998).

19. Maini, R. N., William St. Clair, E., Breedveld, F., Furst, D., Kalden, J., Weisman, M., Smolen, J., Emery, P., Harriman, G., Feldmann, M., Lipsky, P. Randomised phase III trial of infliximab (chimeric anti-TNFα monoclonal antibody) versus placebo in rheumatoid arthritis patients receiving concomitant methotrexate. *Lancet* 354, 1932-1940.

20. Rankin, E. C., Choy, E.H.S., Kassimos, D., Soowith, M., Kingsley, G., Isenberg, D.A. and Panayi, G.S. A double blind, placebo-controlled, ascending dose trial of the

recombinant humanised anti-TNFα antibody CDP571 in patients with rheumatoid arthritis (RA): A preliminary report. *Arthritis Rheum* 37, S295 (1994).

21. Moreland, L.W., Baumgartner, S.W., Schiff, M.H., Tindall, E.A., Fleischmann, R.M., Weaver, A.L., Ettlinger, R.E., Cohen, S., Koopman, W.J., Mohler, K., Widmer, M.B. and Blosch, C.M. Treatment of rheumatoid arthritis with a recombinant human tumor necrosis factor receptor (p75)-Fc fusion protein. *New England Journal of Medicine* 337, 141-147 (1997).

22. Huang, S., Endo, R. I. and Nemerow, G. R. Upregulation of integrins alpha v beta 3 and alpha v beta 5 on human monocytes and T lymphocytes facilitates adenovirus-mediated gene delivery. *J Virol* 69, 2257-2263 (1995).

23. Haddada, H., Lopez, M., Martinache, C., Ragot, T., Abina, M. A. and Perricaudet, M. Efficient adenovirus-mediated gene transfer into human blood monocyte derived macrophages. *Biochem Biophys Res Commun* 195, 1174-1183 (1993).

24. Foxwell, B. M. J. Brown, K., Bondeson, J., Clarke, C. de Martin, R., Brennan, F.M. and Feldmann, M.. Efficient adenoviral infection with IκBα reveals that TNFα production in rheumatoid arthritis is NF-κB dependent. *Proc Natl Acad Sci USA* 95, 8211-8215 (1998).

25. Bondeson, J., Brown, K. A., Brennan, F. M., Foxwell, B. M. J. and Feldmann, M. Selective regulation of cytokine induction by adenoviral gene transfer of IκBα into human macrophages: LPS-induced but not zymosan induced, pro-inflammatory cytokines are inhibited, but IL-10 is NF-κb independent. *J Immunol* 162, 2939-2945 (1999).

26. Hayes, A. L., Smith, C., Foxwell, B. M. and Brennan, F. M. CD45-induced tumor necrosis factor alpha production in monocytes is phosphatidylinositol 3-kinase-dependent and nuclear factor-kappa B- independent. *J Biol Chem* 274, 33455-61 (1999)

27. Arner, E. C., Pratta, M.A., Decicco, C.P., Xue, C.B., Newton, R.C., Trzaskos, J.M., Magolda, R.L. and Tororella, M.D. Aggrecanase. A target for the design of inhibitors of cartilage degradation. *Ann NY Acad Sci* 878, 92-107 (1999).

28. Bondeson, J., Foxwell, B., Brennan, F. and Feldmann, M. Defining therapeutic targets by using adenovirus: blocking NF-kappaB inhibits both inflammatory and destructive mechanisms in rheumatoid synovium but spares anti-inflammatory mediators. *Proc Natl Acad Sci USA* 96, 5668-73 (1999).

29. Zandi, E., Rothwarf, D. M., Delhase, M., Hayakawa, M. and Karin, M. The IκB kinase complex (IKK) contains two kinase subunits, IKKα and IKKβ, necessary for IκB phosphorylation and NF-κB activation. *Cell* 91, 243-252 (1997).

30. Regnier, C. H., Song, H.-Y., Gao, X., Goeddel, D. V., Cao, Z. and Rothe, M. Identification and characterization of an IκB kinase. *Cell* 90, 373-383 (1997).

31. Wallach, D., Varfolomeev, E. E., Malinin, N. L., Goltsev, Y. V., Kovalenko, A.V. and Boldin, M. P. Tumor necrosis factor receptor and Fas signaling mechanisms. *Annu Rev Immunol* 17, 331 -367 (1999).

32. Makarov, S. S., Johnston, W.N., Olsen, J.C., Watson, J.M., Mondal, K., Rinehart, C. and Haskill, J.S. NF-κB as a target for anti-inflammatory gene therapy: suppression of inflammatory responses in monocytic and stromal cells by stable gene transfer of IκBα cDNA. *Gene Therapy* 4, 846-852 (1997).

33. Falta, M. T. and Kotzin, B. L. in *T Cells in Arthritis* (eds Miossec, P., van den Berg, W. B. and Firestein, G. S.) 201-231 (Birkhauser Verlag, Basel, 1998).

34. Firestein, G. S. and Zvaifler, N. J. How important are T cells in chronic rheumatoid synovitis? *Arthritis Rheum* 33, 768-773 (1990).

35. Miossec, P., van den Berg, W. B. and Firestein, G. S. in *Progress in Inflammation Research* (ed Parnham, M. J.) 236 (Birkhauser, Basel, 1998).

36. Tugwell, P., Bombardier, C., Gent, M., Bennett, K.J., Bensen, W.G. Klinkhoff, A.V., Kraag, G.R., Ludwin, D. and Roberts, R.S. Low-dose cyclosporin versus placebo in patients with rheumatoid arthritis. *Lancet* 335, 1051 - 1055 (1990).

37. Isler, P., Vey, E., Zhang, J. II. and Dayer, J. M. Cell surface glycoproteins expressed on activated human T cells induce production of interleukin-1 beta by monocytic cells: a possible role of CD69. *Eur. Cytokine Netw.* 4, 15-23 (1993).

38. Brennan, F. M., Hayes, A. L., Ciesielski, C. J., Foxwell, B. M. J. and Feldmann, M. Cytokine activated T cells are important in TNFα synthesis in rheumatoid arthritis: P13 kinase and NF-κB pathways discriminate between cytokine and TCR activated T cells. *Nature Medicine*, in press (2000).

39. Unutmaz, D., Pileri, P. and Abrignani, S. Antigen-independent activation of naïve and memory resting T cells by a cytokine combination. *J Exp Med* 180, 1159-1164 (1994).

40. Sebbag, M., Parry, S. L., Brennan, F. M. and Feldmann, M. Cytokine stimulation of T lymphocytes regulates their capacity to induce monocyte production of TNFα but not IL-10: possible relevance to pathophysiology of rheumatoid arthritis. *Eur J Immunol* 27, 624-632 (1997).

INDEX